Digital Technologies: Applications, Window of Opportunity and Challenges in Exercise, Health and Sports

Digital Technologies: Applications, Window of Opportunity and Challenges in Exercise, Health and Sports

Guest Editors

Rodrigo Zacca
Flávio Antônio de Souza Castro
Rui Miguel Simões de Azevedo

MDPI

Basel • Beijing • Wuhan • Barcelona • Belgrade • Novi Sad • Cluj • Manchester

Guest Editors
Rodrigo Zacca
Research Center in Physical Activity, Health and Leisure (CIAFEL)
University of Porto
Porto
Portugal

Flávio Antônio de Souza Castro
Aquatic Sports Research Group (GPEA) Universidade Federal do Rio Grande do Sul
Porto Alegre
Brazil

Rui Miguel Simões de Azevedo
Institute for Health and Bioeconomy
University Institute of Health Sciences—CESPU
Carapinheira
Portugal

Editorial Office
MDPI AG
Grosspeteranlage 5
4052 Basel, Switzerland

This is a reprint of articles from the Special Issue published online in the open access journal *Sports* (ISSN 2075-4663) (available at: www.mdpi.com/journal/sports/special_issues/Technologies_sports).

For citation purposes, cite each article independently as indicated on the article page online and using the guide below:

Lastname, A.A.; Lastname, B.B. Article Title. *Journal Name* **Year**, *Volume Number*, Page Range.

ISBN 978-3-7258-1986-7 (Hbk)
ISBN 978-3-7258-1985-0 (PDF)
https://doi.org/10.3390/books978-3-7258-1985-0

Contents

Preface

This reprint represents the collaborative efforts of the Guest Editors to explore new and emerging methodologies for delivering innovative solutions in exercise, health, and sports. We are grateful to all the authors of the manuscripts published in this special edition.

Without their contributions, this reprint would not have been possible. The communities of "Sport Sciences", "Public, Environmental, and Occupational Health" and "Physical Therapy, Sports Therapy, and Rehabilitation" extend their thanks to the authors for their valuable input.

We hope you enjoy reading this collection and that you find the insights within both enlightening and inspiring.

Rodrigo Zacca, Flávio Antônio de Souza Castro, and Rui Miguel Simões de Azevedo
Guest Editors

agreement between the length measures from manual versus LiDAR-based assessment were analysed by calculating the coefficient of determination (R^2) and visualizing the data using Bland–Altman plots, with the confidence interval set at 95% limits of agreement [8].

3. Results

The data of all the 22 participants were included in the evaluation. The average thigh and shank lengths of both legs from the manual and the LiDAR-based assessments including statistical results are given in Table 1. Thus, the scans of six participants were either too distorted or contained missing data, thereby unable to extract shank and thigh lengths (i.e., poor scans), and the scans of another five participants only allowed thigh length measures to be derived (i.e., moderate scans). The scans of eleven participants were categorized as good, allowing for the extraction of shank and thigh lengths of both legs from the 3D point clouds. Consequently, this categorization yielded 16 thigh length measurements (from the 'moderate' and 'good' groups, n = 5 + 11) and 11 shank length measurements (from the 'good' group, n = 11). A representative sample of LiDAR scans is given in Figure 1, and the results from the correlation analysis and Bland–Altman plots are shown in Figure 2.

Table 1. Average thigh and shank lengths of each participant from manual assessment compared to LiDAR-based assessment with *p*-values given from Student's paired *t*-test, coefficient of determination (R^2) from correlation analysis, as well as bias, upper and lower limit from the Bland–Altman analysis, respectively. For each participant (i.e., n = 16 thigh, n = 11 shank), the length measures of the left and right leg were combined for statistical analysis.

	Manual	LiDAR	*p*-Value	R^2	Bias	Upper Limit	Lower Limit
Thigh length [cm] (*n* = 16)	45.8 (3.73)	45.0 (3.38)	0.203	0.901	0.21	2.92	−2.50
Shank length [cm] (*n* = 11)	40.3 (3.01)	38.2 (2.14) *	0.001	0.830	0.89	3.17	−1.40

* indicates a significant difference between ISAK and LIDAR-based assessment with $p < 0.05$.

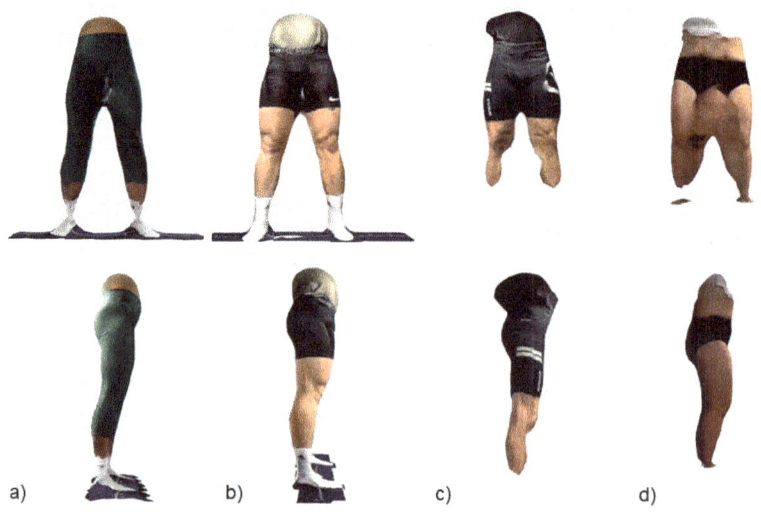

Figure 1. Representative sample of LiDAR scans of the lower limbs from frontal (**top row**) and side view (**bottom row**). (**a**) Good scan of female participants, (**b**) good scan of male participant, (**c**) incomplete scan of male participant, (**d**) incomplete and distorted scan of female participant. Visualisation and stillshots of the 3D point clouds were done using the iOS application Polycam.

sports MDPI

Editorial

Riding the Digital Wave of Exercise, Health, and Sports Training Optimization

Rodrigo Zacca [1,2,*], Flávio Antônio de Souza Castro [3] and Rui M. S. Azevedo [4,5]

1 Research Center in Physical Activity, Health and Leisure (CIAFEL), Faculty of Sports, University of Porto (FADEUP), 4099-002 Porto, Portugal
2 Laboratory for Integrative and Translational Research in Population Health (ITR), 4050-600 Porto, Portugal
3 Aquatic Sports Research Group (GPEA), School of Physical Education, Physiotherapy and Dance, Universidade Federal do Rio Grande do Sul, Porto Alegre 90690-200, Brazil; souza.castro@ufrgs.br
4 Associate Laboratory i4HB—Institute for Health and Bioeconomy, University Institute of Health Sciences—CESPU, 4585-116 Gandra, Portugal
5 UCIBIO—Applied Molecular Biosciences Unit, Forensics and Biomedical Sciences Research Laboratory, University Institute of Health Sciences (1H-TOXRUN, IUCS-CESPU), 4585-116 Gandra, Portugal
* Correspondence: rzacca@fade.up.pt

check for updates

Citation: Zacca, R.; Castro, F.A.d.S.; Azevedo, R.M.S. Riding the Digital Wave of Exercise, Health, and Sports Training Optimization. *Sports* **2024**, *12*, 203. https://doi.org/10.3390/sports12080203

Received: 8 July 2024
Revised: 12 July 2024
Accepted: 16 July 2024
Published: 25 July 2024

The digital era is opening countless possibilities in "Sport Sciences"; "Public, Environmental, and Occupational Health"; and "Physical Therapy, Sports Therapy, and Rehabilitation" areas [1]. Our Special Issue (Digital Technologies: Applications, Window of Opportunity and Challenges in Exercise, Health and Sports) underlines emerging methodologies, providing some novel solutions for exercise, health, sports, coaches, athletes, and the general population [2–5]. In fact, digital technologies offer huge potential for development in all these domains. Emerging technologies, such as artificial intelligence, blockchain, and the Internet of Things, are prompting us to rethink approaches to exercise, health, and sports training [1,2,5]. As we embrace these innovations, it is important for the "Sport Sciences", "Public, Environmental, and Occupational Health", and "Physical Therapy, Sports Therapy and Rehabilitation" communities to efficiently explore this digital wave. From advanced personalization of exercise and training programs to the use of augmented reality in physical rehabilitation, each innovation opens new gaps to improve fitness, athletic performance, and general health. By navigating this new digital frontier with competence and intelligence, we are not only poised to improve athlete outcomes and the quality of life of the general public, but we are also ready to positively transform the present and future of these areas.

Filtered evidence like systematic reviews and meta-analyses synthesize the literature of RCTs, cohort studies, and, on occasion, case reports. The effort involved demands extensive mental, philosophical, and physical effort from researchers, as well as a monetary investment. Twenty-eight (n = 28) manuscripts were submitted for consideration for our Special Issue. All submitted manuscripts underwent peer review, evaluating the originality and quality of the research, methodological rigor, clarity and relevance of the findings and their applications. Ultimately, the following ten (n = 10) papers were accepted for this Special Issue (one communication, eight articles, and one review):

Communication

(1) Oberhofer, K.; Knopfli, C.; Achermann, B.; Lorenzetti, S.R. Feasibility of Using Laser Imaging Detection and Ranging Technology for Contactless 3D Body Scanning and Anthropometric Assessment of Athletes. *Sports* **2024**, *12*, 92. https://doi.org/10.3390/sports12040092.

Articles

(1) Patti, A.; Giustino, V.; Messina, G.; Figlioli, F.; Cataldi, S.; Poli, L.; Belmonte, G.; Valenza, A.; Amato, A.; Thomas, E.; et al. Effects of Cycling on Spine: A Case–Control Study Using a 3D Scanning Method. *Sports* **2023**, *11*, 227. https://doi.org/10.3390/sports11110227.

(2) Nagovitsyn, R.S.; Valeeva, R.A.; Latypova, L.A. Artificial Intelligence Program for Predicting Wrestlers' Sports Performances. *Sports* **2023**, *11*, 196. https://doi.org/10.3390/sports11100196.

(3) Noteboom, L.; Nijs, A.; Beek, P.J.; van der Helm, F.C.T.; Hoozemans, M.J.M. A Muscle Load Feedback Application for Strength Training: A Proof-of-Concept Study. *Sports* **2023**, *11*, 170. https://doi.org/10.3390/sports11090170.

(4) Achermann, B.; Oberhofer, K.; Ferguson, S.J.; Lorenzetti, S.R. Velocity-Based Strength Training: The Validity and Personal Monitoring of Barbell Velocity with the Apple Watch. *Sports* **2023**, *11*, 125. https://doi.org/10.3390/sports11070125.

(5) Cabarkapa, D.V.; Cabarkapa, D.; Philipp, N.M.; Fry, A.C. Impact of the Anatomical Accelerometer Placement on Vertical Jump Performance Characteristics. *Sports* **2023**, *11*, 92. https://doi.org/10.3390/sports11040092.

(6) Luginbühl, M.; Gross, M.; Lorenzetti, S.; Graf, D.; Bünner, M.J. Identification of Optimal Movement Patterns for Energy Pumping. *Sports* **2023**, *11*, 31. https://doi.org/10.3390/sports11020031.

(7) Daveri, M.; Fusco, A.; Cortis, C.; Mascherini, G. Effectiveness of Different Modalities of Remote Online Training in Young Healthy Males. *Sports* **2022**, *10*, 170. https://doi.org/10.3390/sports10110170.

(8) Touloudi, E.; Hassandra, M.; Galanis, E.; Goudas, M.; Theodorakis, Y. Applicability of an Immersive Virtual Reality Exercise Training System for Office Workers during Working Hours. *Sports* **2022**, *10*, 104. https://doi.org/10.3390/sports10070104.

Review

(1) Rossoni, A.; Vecchiato, M.; Brugin, E.; Tranchita, E.; Adami, P.E.; Bartesaghi, M.; Cavarretta, E.; Palermi, S. The eSports Medicine: Pre-Participation Screening and Injuries Management—An Update. *Sports* **2023**, *11*, 34. https://doi.org/10.3390/sports11020034.

These manuscripts were authored by researchers from various countries around the globe, including France, Greece, Italy, Monaco, Poland, Russia, Spain, Switzerland, the Netherlands, and the United States. As editors, we are very grateful for the trust placed in us and are confident that upon reading this Special Issue, you will gain a fresh perspective on "Sport Sciences"; "Public, Environmental, and Occupational Health"; and "Physical Therapy, Sports Therapy, and Rehabilitation".

Author Contributions: Conceptualization, R.Z., F.A.d.S.C., and R.M.S.A.; methodology, R.Z., F.A.d.S.C., and R.M.S.A.; formal analysis, R.Z., F.A.d.S.C., and R.M.S.A.; investigation, R.Z., F.A.d.S.C., and R.M.S.A.; writing—original draft preparation, R.Z., F.A.d.S.C., and R.M.S.A.; writing—review and editing, R.Z., F.A.d.S.C., and R.M.S.A.; visualization, R.Z., F.A.d.S.C., and R.M.S.A. All authors have read and agreed to the published version of the manuscript.

Funding: Rodrigo Zacca is supported by *The Research Center in Physical Activity, Health and Leisure* (CIAFEL), *Faculty of Sport, University of Porto* (FADEUP), which is part of the *Laboratory for Integrative and Translational Research in Population Health* (ITR); both are funded by the *Fundação Para a Ciência e Tecnologia* (FCT; grants UIDB/00617/2020 https://doi.org/10.54499/UIDB/00617/2020; UIDP/00617/2020 https://doi.org/10.54499/UIDP/00617/2020 and LA/P/0064/2020, respectively).

Conflicts of Interest: The authors declare no conflicts of interest.

References

1. Louis, J.; Bennett, S.; Owens, D.J.; Tiollier, E.; Brocherie, F.; Carneiro, M.A.S.; Nunes, P.R.P.; Costa, B.; Castro-e-Souza, P.; Lima, L.A.; et al. Commentaries on Viewpoint: Hoping for the best, prepared for the worst: Can we perform remote data collection in sport sciences? *J. Appl. Physiol.* **2022**, *133*, 1433–1440. [CrossRef] [PubMed]
2. Tehrani, F.; Teymourian, H.; Wuerstle, B.; Kavner, J.; Patel, R.; Furmidge, A.; Aghavali, R.; Hosseini-Toudeshki, H.; Brown, C.; Zhang, F.; et al. An integrated wearable microneedle array for the continuous monitoring of multiple biomarkers in interstitial fluid. *Nat. Biomed. Eng.* **2022**, *6*, 1214–1224. [CrossRef] [PubMed]
3. Hettiarachchi, I.T.; Hanoun, S.; Nahavandi, D.; Nahavandi, S. Validation of Polar OH1 optical heart rate sensor for moderate and high intensity physical activities. *PLoS ONE* **2019**, *14*, e0217288. [CrossRef] [PubMed]
4. Zacca, R.; Castro, F.A.S.; Monteiro, A.S.M.; Pyne, D.B.; Vilas-Boas, J.P.; Fernandes, R.J.P. Swimming with the COSMED AquaTrainer and K5 Wearable Metabolic System in Breath-by-Breath Mode: Accuracy, Precision, and Repeatability. *Int. J. Sports Physiol. Perform.* **2023**, *18*, 1152–1160. [CrossRef] [PubMed]
5. Van Hooren, B.; Plasqui, G.; Meijer, K. The Effect of Wearable-Based Real-Time Feedback on Running Injuries and Running Performance: A Randomized Controlled Trial. *Am. J. Sports Med.* **2024**, *52*, 750–765. [CrossRef] [PubMed]

sports MDPI

Communication

Feasibility of Using Laser Imaging Detection and Ranging Technology for Contactless 3D Body Scanning and Anthropometric Assessment of Athletes

Katja Oberhofer [1], Céline Knopfli [1,2], Basil Achermann [1,2] and Silvio R. Lorenzetti [1,2,*]

[1] Section Performance Sport, Swiss Federal Institute of Sport Magglingen (SFISM), Hauptstrasse 247, 2532 Magglingen, Switzerland; katjaoberhofer@outlook.com (K.O.); celine.knopfli@hest.ethz.ch (C.K.); basil.achermann@baspo.admin.ch (B.A.)

[2] Department of Health Sciences and Technology (HEST), ETH Zurich, LFW C13.2, Universitätsstrasse 2, 8092 Zurich, Switzerland

* Correspondence: sl@ethz.ch

Abstract: The scope of this pilot study was to assess the feasibility of using the laser imaging detection and ranging (LiDAR) technology for contactless 3D body scanning of sports athletes and deriving anthropometric measurements of the lower limbs using available software. An Apple iPad Pro 3rd Generation with embedded LiDAR technology in combination with the iOS application Polycam were used. The effects of stance width, clothing, background, lighting, scan distance and measurement speed were initially assessed by scanning the lower limbs of one test person multiple times. Following these tests, the lower limbs of 12 male and 10 female participants were scanned. The resulting scans of the lower limbs were complete for half of the participants and categorized as good in quality, while the other scans were either distorted or presented missing data around the shank and/or the thigh. Bland–Altman plots between the LiDAR-based and manual anthropometric measures showed good agreement, with the coefficient of determination from correlation analysis being $R^2 = 0.901$ for thigh length and $R^2 = 0.830$ for shank length, respectively. The outcome of this pilot study is considered promising, and a further refinement of the proposed scanning protocol and advancement of available software for 3D reconstruction are recommended to exploit the full potential of the LiDAR technology for the contactless anthropometric assessment of athletes.

Keywords: anthropometry; LiDAR; 3D shape analysis; body sizing; strength training; sports science

check for updates

Citation: Oberhofer, K.; Knopfli, C.; Achermann, B.; Lorenzetti, S.R. Feasibility of Using Laser Imaging Detection and Ranging Technology for Contactless 3D Body Scanning and Anthropometric Assessment of Athletes. *Sports* **2024**, *12*, 92. https://doi.org/10.3390/sports12040092

Academic Editors: Flávio Antônio de Souza Castro, Rui Miguel Simões de Azevedo and Rodrigo Zacca

Received: 12 October 2023
Revised: 17 November 2023
Accepted: 22 March 2024
Published: 26 March 2024

1. Introduction

Subject-specific anthropometric measurements are needed in a broad context across ergonomics, engineering, design research, health and sports sciences. Therefore, subject-specific body measurements are traditionally obtained by hand according to the standards of the International Society for the Advancement of Kinanthropometry (ISAK) and used for, e.g., designing new customer goods, assessing patient characteristics or monitoring training progress [1].

In recent years, automatic 3D scanners have provided new means to capture body surface data of individual subjects, and these are contactless, with high repeatability and speed [2]. Particularly, a new light detection and ranging (LiDAR) sensor for depth sensing was introduced in 2020 by Apple (Apple Inc., Cupertino, CA, USA) into their high-end mobile devices, which has opened the way for convenient 3D scanning outside the laboratory. The LiDAR technology works by emitting arrays of infrared light pulses from a series of transmitters into the environment, which are reflected from the surface of the target object and re-captured by integrated photodetector sensors. The sensors detect the frequency of the reflected light, which is then used to calculate travel time and distance to the target surface [2].

The Apple LiDAR technology has been adopted for, e.g., forensic 3D documentation [3], large animal assessments in agriculture [4] and to estimate tree diameter in the context of forest management [5]. Furthermore, 3D scanning using commercial mobile devices has proven useful for preoperative and postoperative analysis of facial structures by plastic surgeons [6], as well as the estimation of body segment parameters for biomechanical analysis [7]. Yet, the potential of the Apple LiDAR technology for 3D body scanning and anthropometric assessment of sports athletes has not yet been demonstrated.

The goal of this pilot study was to assess the feasibility of using the Apple LiDAR technology for contactless anthropometric measurements of strength-training athletes outside the dedicated laboratory and to assess the feasibility of extracting anthropometric measures from the 3D data using available iOS software (Version 15.5). It was hypothesized that the LiDAR technology allows for the contactless measurement of shank and thigh length based on 3D body surface scanning in a training-specific setting with manual measurements, according to ISAK standards as reference values.

2. Materials and Methods

Ethical approval for this study was given by the regional ethics committee (Kantonale Ethikkommission Bern, Nr: 2021-00403). A total of 22 healthy, recreationally active subjects (n = 12 M/10 F, age = 29 ± 4.7, height = 1.64 ± 0.38 m; body mass = 76 ± 12 kg) gave written informed consent to participate in this study. An Apple iPad Pro 3rd Generation with embedded LiDAR technology was used in combination with the iOS application Polycam for 3D scanning, visualization and analysis (https://poly.cam, accessed on 1 September 2022).

The effect of leg distance, clothing, background, lighting, scan distance and measurement speed were initially assessed by scanning the lower limbs of one test person multiple times. Based on this initial testing, the measurement protocol for scanning the lower limbs was defined as follows: (1) uniform background and lighting, (2) participant wearing tight, single-coloured clothing or only presenting naked skin, (3) participant standing in a T-position with standardized leg distance of 25–30% body height, (4) examiner moving at constant, moderate speed on a circular path around the target, (5) keeping the iPad as stable as possible, perpendicular to the plane of motion and (6) keeping a constant distance of 50–100 cm from the target.

Each subject was scanned according to the above protocol, whereby all scans were performed by one examiner. The same examiner also performed the test scans in order to familiarize herself with the technology. Additionally, anthropometric data of all participants, including size, weight, thigh and shank lengths and circumferences, were manually measured by a trained practitioner according to ISAK standards. Each measure was taken twice to be averaged. Data acquisition was not randomised. For all participants, the LiDAR scan was firstly obtained, followed by manual anthropometric measurements.

The 3D models from LiDAR scanning were visually assessed and analysed using the iOS application Polycam. Measurements of thigh and shank length were extracted as study outcome parameters from the 3D models using the integrated linear measuring tool. Thus, 3D models were categorized into 'poor', 'moderate' and 'good' depending on the completeness of body surface data. Particularly, 3D models were categorized as 'poor' if the extraction of thigh and shank lengths was not possible and as 'moderate' if only thigh or shank length could be individually extracted.

The length measures from the left and right leg of all participants were combined and statistically compared between the manual and the LiDAR-based data. The comparison was limited to thigh and shank length measurements due to the constraints of the Polycam software (Version 3.2.7), which only allowed for linear measurements to be extracted. Prior to statistical analysis, data were checked for normal distribution using the Shapiro–Wilk test. Student's paired t-tests were then used to determine whether the differences in lengths measures between the manual and the LiDAR-based measures were statistically significant, with the level of significance set at $p < 0.05$. Furthermore, the correlation and

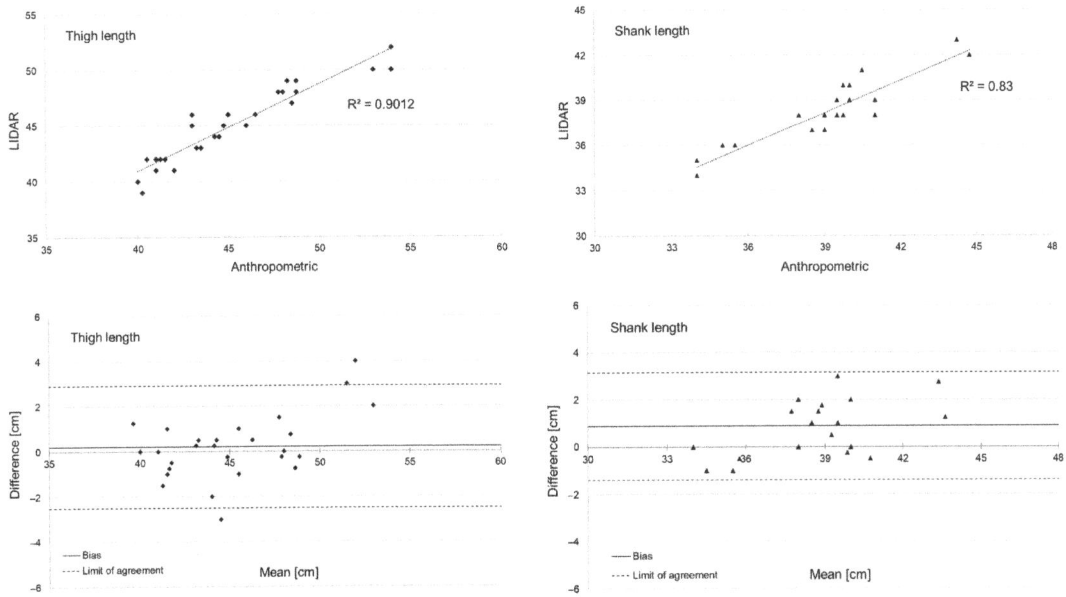

Figure 2. Correlation (**top**) with coefficient of determination (R^2) and Bland–Altman plots (**bottom**) with the confidence interval at 95% limits of agreement between LiDAR-based and manual anthropometric measures of thigh length (**left**) and shank length (**right**). Length measures from the left and right leg of each participant (i.e., $n = 16$ thigh, $n = 11$ shank) were taken into account.

4. Discussion

Given the growing popularity of LiDAR technology as a consumer electronic device, the goal of this pilot study was to provide guidelines and praxis-oriented insights into the potential of the Apple LiDAR technology for convenient anthropometric assessment in a sport-specific setting. Based on initial testing, the measurement protocol was defined to ensure uniform background and lighting, with the examiner moving at a constant moderate speed around the subject and keeping the iPad stable and perpendicular to the plane of motion at a constant distance to the target. Nevertheless, half of the resulting scans were only moderate or poor in quality, with distortions or missing data especially between the legs and closer to the floor (Figure 1).

Inconsistent lighting between the legs, as well as around the shank and ankle close to the floor, may have contributed to the poor 3D reconstruction in these areas. Unfortunately, no decisive conclusion could be drawn regarding the best choice of garment, colour and/or bare skin to improve scan quality. In similar work of facial scanning, it was also found that areas with inconsistent lighting and increased specular reflectivity (e.g., nose and chin) led to higher inaccuracies [6]. Thereby, the influence of skin type on scan outcome was also inconsistent in previous work [6]. Further experiments with additional adjustments to the present scan protocol are thus highly recommended, including other software packages for reconstruction.

The quality of scans using LiDAR technology is largely dependent on the 3D reconstruction capability of the chosen software. In the present work, the iOS application Polycam was used for 3D scanning and data visualization. Unfortunately, the integrated software tool only allowed for the extraction of linear measurements from the 3D point cloud, i.e., shank and thigh lengths but not circumferences. There are a few fitness-specific stationary 3D scanners available, as well as mobile applications to estimate body dimensions based on RGB images from different views [2]. Yet, to the authors knowledge, there is no software available for anthropometric assessment based on LiDAR data. Further software development is highly encouraged and may likely help to improve analysis results.

This pilot study was part of a larger study to improve the safety and efficiency of strength training by means of mobile technology [9,10]. These results indicate a degree of reliability (Table 1), but further validation studies with a repeated measurement design and larger sample sizes are needed to draw decisive conclusions regarding the validity and reliability of LiDAR technology for contactless anthropometric assessment. The potential advantage of using 3D scanning technology compared to manual anthropometric assessment is the measurement speed, with the duration of a scan taking less than 1 min, as well as the possibility for the layperson to obtain an accurate measurement of body dimensions without prior training. Additionally, larger and more diverse body scanning datasets may become publicly available, with advances in deep learning algorithms and optimization techniques as well for the improved monitoring of physical training and rehabilitation progress.

5. Conclusions

Advancements in LiDAR technology, as embedded in mobile devices, are opening the doors for convenient 3D body surface scanning and anthropometric assessment in a sport-specific setting. Despite challenges with inconsistent lighting across body parts and remaining software limitations, the outcomes of this pilot study are considered promising. Further advancements of the proposed scanning protocol and available software for 3D reconstruction are highly recommended to exploit the full potential of the LiDAR technology. For validation purposes, future studies should consider a repeated measurement design with larger sample sizes to substantiate the present preliminary results with scientific rigor. The ability to conveniently assess subject-specific body dimensions using mobile devices outside the dedicated laboratory is expected to help in the monitoring of training and rehabilitation outcomes to the benefit of athletes and patients alike.

Author Contributions: Conceptualization, investigation, formal analysis, writing—original draft, K.O.; data acquisition, visualization, writing—review and editing, C.K.; software, data curation, writing—review and editing, B.A.; funding acquisition, resources, project administration, writing—review and editing, S.R.L. All authors have read and agreed to the published version of the manuscript.

Funding: This research was funded by the Swiss National Science Foundation (SNF), Nr 200021_192289/1.

Institutional Review Board Statement: This study was conducted in accordance with the Declaration of Helsinki and approved by the Kantonale Ethikkommission für die Forschung (KEK) Bern (Project-ID 2021-00403).

Informed Consent Statement: Signed informed consent was obtained from all subjects involved in the study.

Data Availability Statement: Data from this study are unavailable due to ethical restrictions.

Acknowledgments: We would like to thank all the participants and the resources provided by Cross Fit Gleis 10 Winterthur.

Conflicts of Interest: The authors declare no conflicts of interest. The funders had no role in the design of the study; in the collection, analyses, or interpretation of data; in the writing of the manuscript; or in the decision to publish the results.

References

1. Norton, K. Standards for Athropometry Assessment. In *Kinanthropometry and Exercise Physiology*; Routledge: London, UK, 2018; pp. 68–137. ISBN 978-1-315-38566-2.
2. Bartol, K.; Bojanić, D.; Petković, T.; Pribanić, T. A Review of Body Measurement Using 3D Scanning. *IEEE Access* **2021**, *9*, 67281–67301. [CrossRef]

3. Kottner, S.; Thali, M.J.; Gascho, D. Using the IPhone's LiDAR Technology to Capture 3D Forensic Data at Crime and Crash Scenes. *Forensic Imaging* **2023**, *32*, 200535. [CrossRef]
4. Huang, L.; Li, S.; Zhu, A.; Fan, X.; Zhang, C.; Wang, H. Non-Contact Body Measurement for Qinchuan Cattle with LiDAR Sensor. *Sensors* **2018**, *18*, 3014. [CrossRef]
5. Wang, F.; Heenkenda, M.K.; Freeburn, J.T. Estimating Tree Diameter at Breast Height (DBH) Using an IPad Pro LiDAR Sensor. *Remote Sens. Lett.* **2022**, *13*, 568–578. [CrossRef]
6. Nightingale, R.C.; Ross, M.T.; Allenby, M.C.; Woodruff, M.A.; Powell, S.K. A Method for Economical Smartphone-Based Clinical 3D Facial Scanning. *J. Prosthodont. Off. J. Am. Coll. Prosthodont.* **2020**, *29*, 818–825. [CrossRef] [PubMed]
7. Kudzia, P.; Jackson, E.; Dumas, G. Estimating Body Segment Parameters from Three-Dimensional Human Body Scans. *PLoS ONE* **2022**, *17*, e0262296. [CrossRef] [PubMed]
8. Bland, J.M.; Altman, D.G. Statistical Methods for Assessing Agreement between Two Methods of Clinical Measurement. *Lancet* **1986**, *1*, 307–310. [CrossRef]
9. Knopfli, C.; Achermann, B.; Oberhofer, K.; Lorenzetti, S.R. First Insights in the Relationship between Lower Limb Anatomy and Back Squat Performance in Resistance-Trained Males and Females. *Bioengineering* **2023**, *10*, 865. [CrossRef] [PubMed]
10. Oberhofer, K.; Lorenzetti, S.; Mithraratne, K. Host Mesh Fitting of a Generic Musculoskeletal Model of the Lower Limbs to Subject-Specific Body Surface Data: A Validation Study. *Appl. Bionics Biomech.* **2019**, *2019*, 8381351. [CrossRef] [PubMed]

sports

MDPI

Article

Effects of Cycling on Spine: A Case–Control Study Using a 3D Scanning Method

Antonino Patti [1], Valerio Giustino [1], Giuseppe Messina [2,3,*], Flavia Figlioli [1,*], Stefania Cataldi [4], Luca Poli [4], Giacomo Belmonte [1], Alessandro Valenza [1], Alessandra Amato [5], Ewan Thomas [1], Izabela Rutkowska [6], Paula Esteban-García [7], Antonio Palma [1] and Antonino Bianco [1]

1 Sport and Exercise Sciences Research Unit, Department of Psychology, Educational Science and Human Movement, University of Palermo, 90144 Palermo, Italy; antonino.patti01@unipa.it (A.P.); valerio.giustino@unipa.it (V.G.); ewan.thomas@unipa.it (E.T.); antonino.bianco@unipa.it (A.B.)
2 Department of Human Sciences and Promotion of the Quality of Life, San Raffaele University, 00133 Rome, Italy
3 PLab Research Institute, 90131 Palermo, Italy
4 Department of Basic Medical Sciences, Neuroscience and Sense Organs, University of Study of Bari, 70124 Bari, Italy; luca.poli@uniba.it (L.P.)
5 Department of Biomedical and Biotechnological Sciences, Section of Anatomy, Histology and Movement Science, University of Catania, 95123 Catania, Italy; alessandra.amato@unict.it
6 Faculty of Rehabilitation, University of Physical Education in Warsaw, 00-968 Warsaw, Poland
7 Performance and Sport Rehabilitation Laboratory, PerlaSport Group, Faculty of Physical Activity and Sport Science, University of Castilla la Mancha, 45071 Toledo, Spain
* Correspondence: giuseppe.messina@uniroma5.it (G.M.); flavia.figlioli@unipa.it (F.F.)

Abstract: Background: Few studies have investigated the effects of adopting a specific and prolonged posture on cyclists. This study aimed to evaluate the upright spine in a sample of recreational cyclists and compare it with a sample of non-cyclists, though still athletes, through a 3D scanning method. Methods: Forty-eight participants were enrolled in this observational study. The sample consisted of 25 cyclists for the cycling group and 23 non-cyclist athletes for the control group. The Spine3D device (Sensor Medica, Guidonia Montecelio, Rome, Italy) was used to evaluate the spine of the participants in both groups. Results: The results showed significantly greater spine inclination in the cycling group compared to the control group ($p < 0.01$). Furthermore, there was a significant decrease in lumbar lordosis in the cycling group compared to the control group ($p < 0.01$). Conclusions: This case–control study raises the possibility that the onset of lower back pain in cyclists may be due to a reduction in lumbar lordosis. Furthermore, this study demonstrated that the Spine3D device can be used in sports to monitor the spine of athletes to prevent and reduce musculoskeletal deficits.

Keywords: bicycle; bike; cycling; cyclists; biomechanics; posture; spine; lower back pain

Citation: Patti, A.; Giustino, V.; Messina, G.; Figlioli, F.; Cataldi, S.; Poli, L.; Belmonte, G.; Valenza, A.; Amato, A.; Thomas, E.; et al. Effects of Cycling on Spine: A Case–Control Study Using a 3D Scanning Method. *Sports* **2023**, *11*, 227. https://doi.org/10.3390/sports11110227

Academic Editors: Rodrigo Zacca, Flávio Antônio de Souza Castro and Rui Miguel Simões de Azevedo

Received: 6 October 2023
Revised: 6 November 2023
Accepted: 14 November 2023
Published: 15 November 2023

1. Introduction

The literature on the biomechanics of the technical gestures of athletes in different sports is extensive, and some studies have shown that specific positions could influence the curves of the spine [1–4].

The practice of recreational cycling is increasing, and the literature has demonstrated the multiple health benefits of cycling, as well as indoor cycling has been the subject of interest in recent years for the same purposes [5–7]. However, cycling is a sport characterized by a close relationship between the human body and a mechanical means, namely the bicycle. Cyclists must adopt a peculiar position on the bike and, for this reason, their spine must adopt a specific position to achieve the best aerodynamics [8]. The stance of cyclists is strongly influenced by the three points of contact: the saddle, the handlebars, and the pedals. Intense training sessions, resulting in prolonged sitting times, can lead to adaptations of the spine and cause increased stress on the spine [9–11]. The spine of cyclists undergoes a three-dimensional reorganization during pedaling. In particular, alterations in the lumbar lordotic curve can be observed while cyclists pedaling. In the pursuit of the optimal aerodynamics, cyclists tend to increase pelvic inclination and lumbar flexion [12]. Muyor et al. showed an increase in thoracic kyphosis, a decrease in lumbar lordosis, and a greater pelvic inclination in a group of cyclists [13]. The same authors in a previous study reported comparable findings identifying a high prevalence of thoracic hyperkyphosis in the upright position and a kyphotic posture of the lumbar curve in cyclists [14].

Previous research has shown a relationship between cyclists' lumbar posture and Lower Back Pain (LBP) [9]. Numerous studies have demonstrated a connection between LBP and recurrent forward bending and extended periods of sitting with the lumbar spine in a flexed posture [15–17]. Furthermore, this spinal flexion has been linked to elevated pressures on the intervertebral discs [18]. The literature indicates that cyclists might be susceptible to "mechanical creep" [19]. This biomechanical phenomenon pertains to the gradual deformation or a change in the strain of the ligament tissue when subjected to a sustained load [20,21]. Some studies propose that mechanical creep could occur in the ligaments of the lumbar spine during long periods of sitting in a flexed position on a bike [19].

On the other hand, overuse injuries occur in those who ride bicycles regularly and can be caused by inadequate preparation or an incorrect pedaling position [22]. In a recent article, Cyr reported the need for a spine assessment and recommended a bike fit assessment to counteract neck pain in cyclists [23]. Salai et al. described that 30–70% of cyclists suffer from cervical, dorsal, or lumbar back pain [15]. LBP is common in cycling, but the incidence of neck pain has been poorly investigated.

Many studies have evaluated the position of cyclists on the bicycle, but a small number of studies have been conducted to assess the upright posture of cyclists. Cycling could influence, in the long-term, the morphology of the spine and, therefore, the upright posture. The gold standard for longitudinal spine evaluation is two-dimensional (2D) posterior–anterior full-length spine radiography [24]. However, frequent assessments could lead to long-term adverse effects [25,26]. Rasterstereography is a method used to make a stereophotogrammetric back measurement and facilitates clinical practice by examining the spine [27,28]. The literature suggests that the stereophotogrammetric system is a valid method for evaluating spinal curves [29,30]. In a recent study, Marin et al. used a new alternative technological approach based on infrared cameras using LiDAR technology to evaluate the spine. Moreover, this system does not require room darkness for image acquisition [31]. Recently, Roggio et al. demonstrated the utility and conformity of this technology for the evaluation of spinal alterations [32]. Noninvasive screening methods can detect a specific alteration before the individual experiences discomfort or pain. This could make this technology valuable in preventing spinal problems among athletes, particularly non-professional ones who have less oversight.

Considering the limited studies available in the literature on this topic and the preliminary nature of our study, a case–control design appears to be suitable. Hence, this study aimed to assess the upright spine in a sample of recreational cyclists and compare it with a sample of non-cyclists through a 3D scanning method to identify a strategy for preventing athletes' LBP. Furthermore, we explored whether there was a functional relationship between the various sections of the spine with compensatory mechanisms. This exploration aimed to identify potential predictors of deficits or determine the relative importance of specific deficits.

2. Materials and Methods

2.1. Study Design and Procedure

This is a case–control study in which the participants were athletes living in Sicily (Italy). The enrolled participants were divided into a cycling group (Cy-G) composed of cycling athletes and a control group (CG) composed of non-cycling athletes. Written informed consent was obtained from all participants prior to participating in the study. The

STROBE flow chart (Figure 1) was used to ensure that the assessment of the participants of the study was conducted in a clear way [33].

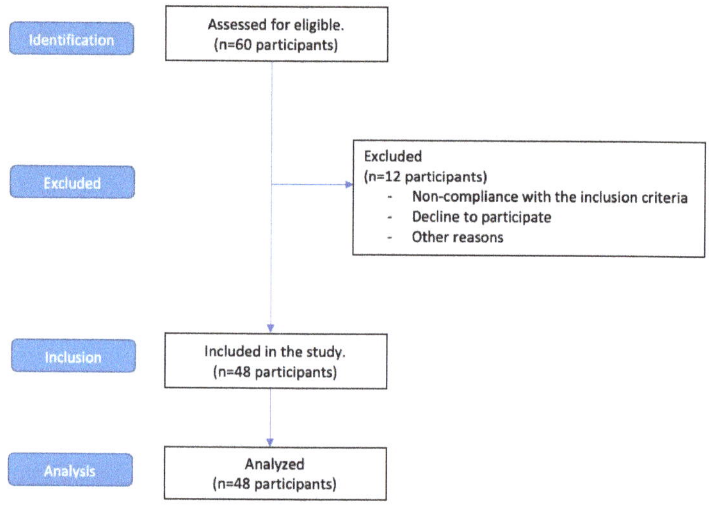

Figure 1. The STROBE flow chart of the study.

The study was carried out in compliance with the principles of the Declaration of Helsinki and approved by the Bioethics Committee of the University of Palermo (Num. 97/2022—Prot. 79743).

2.2. Participants

Sixty participants were recruited for the study, but only forty-eight of them met the inclusion criteria or did not decline to participate in the study, so they were included in this study (Figure 1).

The participants were recruited voluntarily at the Sport and Exercise Sciences Research Centre of the University of Palermo. Specifically, between 4:00 and 6:00 p.m. in a quiet laboratory with a constant temperature (22 °C). The clinical setting is shown in Figure 2.

The participants were interviewed by a researcher, who did not know the purpose of the study, to collect general information on pathologies, allergies, use of drugs, recent surgeries, and sports practiced. According to this information, the exclusion and inclusion criteria were applied. The Cy-G consisted of twenty-five participants and the CG of twenty-three. To be eligible for the Cy-G, participants had to meet the following inclusion criteria: (a) at least four years of cycling practice; (b) at least three days a week of training; (c) no history of spinal pathology. Furthermore, all participants reported engaging in weekly training sessions covering a range of 250 to 350 km at an average cycling speed of 22.0 km/h. This parameter indicates that all the individuals we recruited were recreational cyclists. It is worth noting that the literature suggests that elite athletes typically cycle around 30,000 to 35,000 km annually [34]. To be eligible for the CG, participants had to meet the following inclusion criteria: (a) non-competitive athletes; (b) non-cyclists; (c) not sedentary (at least 150–300 min of moderate-intensity aerobic physical activity per week or at least 75–150 min of vigorous-intensity aerobic physical activity per week) [35]; (d) no history of spinal pathology.

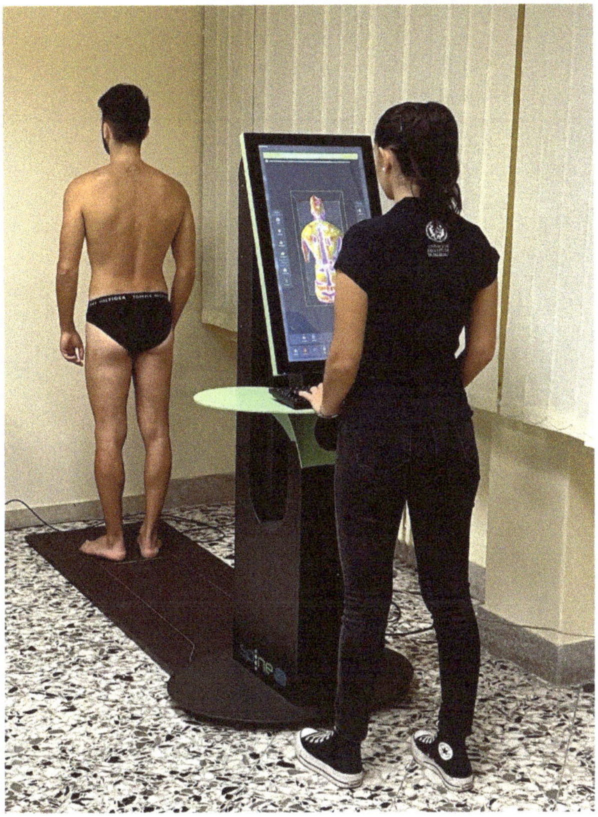

Figure 2. The clinical setting.

In the 24 h prior to the test sessions, none of the participants had performed any exercises. Both test sessions were administered in the same setting. The tests were conducted by the same researcher, an expert in the use of the instrument.

2.3. Instrument

First, the weight and height of each participant were recorded. Body mass was measured using a Seca electronic scale (maximum weight: 300 kg, resolution: 100 g; Seca; Hamburg, Germany). Height was measured using a standard stadiometer (maximum height: 220 cm, resolution: 1 mm).

The Spine3D device with the related software (Sensor Medica, Guidonia Montecelio, Rome, Italy) was used to evaluate the morphology of the spine in the upright posture of the participants in both groups. Spine3D is a non-invasive, three-dimensional optoelectronic detection system for the back [25]. This device allows the 3D acquisition of the shoulders, spine, and pelvis through a marker-less, radiation-free, and non-invasive scanning method. The acquisition is made up of infrared cameras using Time-of-Flight (ToF) depth technology. The technique used is Light Detection and Ranging (LIDAR), which allows for determining the distance of a surface by measuring the time between the emission of the laser and the reception of the reflected light by the receiver.

2.4. Reliability of the Instrument

Rasterstereography is an emerging method that uses Light Detection and Ranging (LiDAR) to evaluate postures that are either abnormal or physiological. When working with a large sample, its excellent intra- and interday dependability makes it a useful technique [32,36]. The reliability of most parameters was excellent. According to Rosner [37], Guidetti et al. [36] indicated that ICCs less than ±0.40 indicate poor reliability; ±0.40–0.75 indicates fair or good reliability; and ±0.75–1.00 indicates excellent reliability. The intraclass correlation coefficient (ICC) and Cronbach's Alpha (Cα) were calculated. The authors evaluated the rasterstereographic system on the spine with and without reflecting markers in order to ascertain its intra- and interday reliability. The greater reliability coefficients for trunk length, kyphotic angle, and lordotic apex were 0.971, 0.963, and 0.958 (ICC) and 0.987, 0.983, and 0.985 (Cα) in the group with markers for intra-, interday, and overall evaluations; in the group without reflective markers, they were 0.978, 0.982, and 0.972 and 0.989, 0.991, and 0.991 for trunk length. The trunk and pelvic torsion values in the markers group were 0.598, 0.515, and 0.534 (ICC) and 0.742, 0.682, and 0.784 (Cα), while the left lateral deviation values in the group without reflective markers were 0.561, 0.537, and 0.461 and 0.731, 0.695, and 0.729. Spine3D has a resolution of 1920 × 1080 pixels and a rate of acquisition of 30 frames per second (fps) [36]. In order to assess the spine, each participant was positioned one meter from Spine3D and with their back facing the camera from the device. Each participant, wearing only undergarments, was asked to stay barefoot in an orthostatic position with their head in a neutral position and with their feet placed side-by-side in a neutral position (Figure 2) [38]. The software captures and processes the image of the back using an automatic identification of the reference points [i.e., prominent vertebra (VP), right and left shoulder (SR and SL), right and left lumbar dimple (DR and DL) and automatically calculates the midpoint between them (DM), the sacrum point (SP)] and traces the morphology of the spine with a three-dimensional rendering (resolution of 1 mm) [31]. These reference points can be edited manually.

2.5. Parameters of the Instrument

The software automatically computes various parameters across three planes: sagittal, frontal, and transversal. For this study, we considered the following parameters.

Sagittal plane parameters:

1. Spine Length (Figure 3a): this is the length of the perpendicular segment from the VP to DM.
2. Spine Inclination (Figure 3a): this measures the angle between the line passing through the VP and DM and the line perpendicular to the transversal plane passing through the DM.
3. Cervical Lordosis (Figure 3b): this parameter represents the distance between VP and the tangent to the kyphotic apex, perpendicular to the transversal plane.
4. Lumbar Lordosis (Figure 3b): this quantifies the distance between the lumbar apex and the tangent to the kyphotic apex, perpendicular to the transversal plane.
5. Kyphotic Angle (Figure 3c): this angle is formed by the tangents to the surface at the cervico-thoracic inversion (ICT) and the thoraco-lumbar inversion (ITL) points.
6. Lordotic Angle (Figure 3c): this angle is formed by the tangents to the surface at ITL and the lumbosacral inversion (ILS) points.

These parameters provide a comprehensive evaluation of spine morphology in the sagittal plane for this study.

Frontal plane parameters:

1. Coronal Imbalance (Figure 4a): this is the distance between the line perpendicular to the transversal plane passing through the VP and the line perpendicular to the transversal plane passing through the DM.
2. Spine Imbalance (Figure 4b): This measures the angle between the line passing through the VP and DM and the line perpendicular to the transversal plane passing through the DM. The instrument evaluates the degrees of inclination, which are positive (indicating an inclination to the right) or negative (indicating an inclination to the left).

3. Shoulder Obliquity (Figure 4a): this parameter quantifies the distance between the line parallel to the transverse plane passing through the SL and the line parallel to the transverse plane passing through the SR.
4. Shoulder Inclination (Figure 4b): this calculates the angle between the line parallel to the transverse plane passing through the SL and the line passing through the SL and SR.
5. Pelvic Obliquity (Figure 4a): this represents the distance between the line parallel to the transverse plane passing through the DL and the line parallel to the transverse plane passing through the DR.
6. Pelvic Inclination (Figure 4b): this measures the angle between the line parallel to the transverse plane passing through the DL and the line passing through the DL and DR.

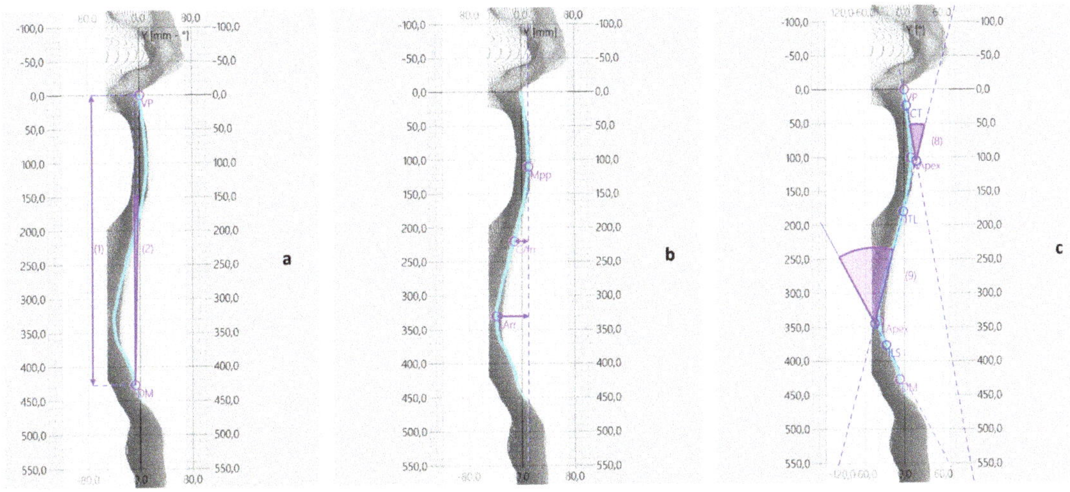

Figure 3. (**a**–**c**): The image shows how the software automatically calculates different parameters.

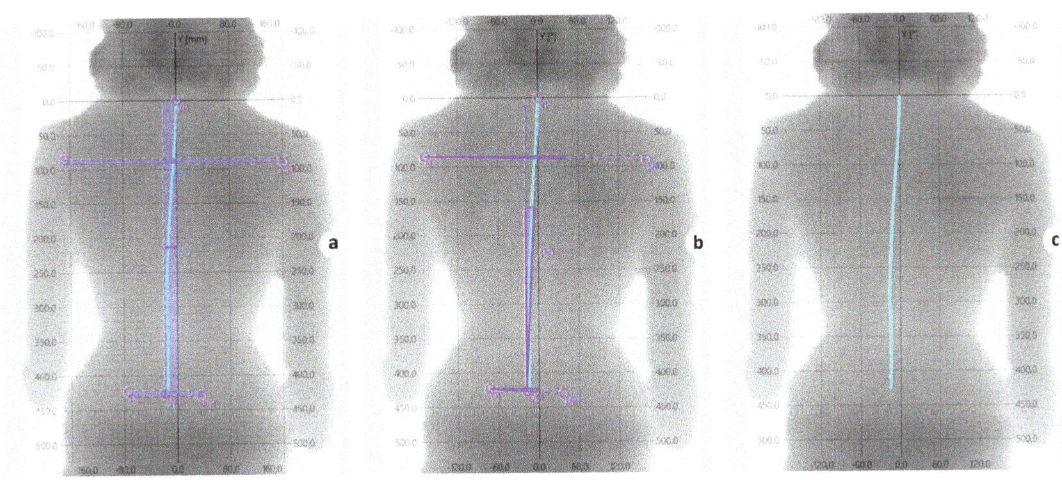

Figure 4. (**a**–**c**): The image shows how the software automatically calculates different parameters.

These parameters provide a comprehensive evaluation of spine morphology in the frontal plane for this study.

2.6. Statistical Analysis

All data were entered into an Excel sheet before being analyzed. Descriptive statistics were reported as the mean ± standard deviation. Shapiro–Wilk's normality test was used to analyze the data distribution. Power analysis of post hoc sample size (α = 0.05) G*Power 3.1.9.2 software (Heinrich Heine University, Düsseldorf, Germany) was used [39].

Differences in spine parameters between the groups (Cy-G and CG) were evaluated using an unpaired *t*-test. For each outcome, Cohen's d was calculated to determine the effect size [40]. Effect size d = 0.2 corresponds to a small effect size; d = 0.5 is medium and d = 0.8 is large [41].

To investigate potential correlations among the spine parameters, Pearson's correlation analysis was utilized. The linear relationship between data is represented by the correlation coefficient "r" and interpreted as follows: positive values indicate a positive linear correlation; negative values signify a negative linear correlation; a value of 0 implies no linear correlation. The closer the value is to 1 or −1, the stronger the linear correlation. Correlation strengths are typically categorized as follows: values below 0.40 or above −0.40 are considered poor; values ranging from 0.40 to 0.59 (or −0.40 to −0.59) are rated as fair; values within the range of 0.60 to 0.74 (or −0.60 to −0.74) are regarded as good; values between 0.75 and 1.00 (or −0.75 and −1.00) are considered excellent [42].

The relationship between Spine Length, Spine Inclination, and Kyphotic Angle was analyzed using linear regressions, including each of these parameters as an independent variable and the Cervical Lordosis as a dependent variable (regression *p*-values and adjusted R-values were calculated). R^2, or the coefficient of determination, is an index that measures the link between the variability of the data and the correctness of the statistical model used [43]. Typically, it ranges from 0 to 1, and a value of 1 means that all of the variances in the dependent variable is explained by the independent variables, indicating a perfect fit of the model to the data [43].

The significance level was set at a *p*-value less than 0.05. Statistical analyses were conducted using Jamovi software (version 2.3.0.0) and GraphPad Prism 8.0.

3. Results

Forty-eight subjects were analyzed (Cy-G: n = 25; CG: n = 23). Shapiro–Wilk's normality test showed a Gaussian distribution of all the parameters. The post hoc sample size power analysis (α = 0.05) showed that, with a total sample size of 48 participants (Cy-G: n = 25; CG: n = 23), we achieved a power of 0.86.

No significant differences in anthropometric characteristics between the groups were found (Cy-G: age = 44.5 ± 12.84 years; body mass = 68.5 ± 8.17 kg; height = 169.6 ± 5.87 cm. CG: age = 41 ± 16.31 years; weight = 64.8 ± 8.03 kg; height = 168.4 ± 6.41 cm).

Table 1 shows descriptive statistics of the spine parameters and differences between the Cy-G and the CG. Significant differences were found in Spine Inclination and Lumbar Lordosis (Table 1). The Spine Inclination and Lumbar Lordosis parameters showed a significant difference compared to the control.

Pearson's analysis showed significant correlations between Spine Length and Lumbar Lordosis (R = 0.30; $p < 0.05$); Spine Length and Cervical Lordosis (R = 0.50; $p < 0.001$); Spine Length and Kyphotic Angle (R = 0.33; $p < 0.05$); Lumbar Lordosis and Spine Inclination (R = −0.36; $p < 0.05$); Lumbar Lordosis and Kyphotic Angle (R = 0.38; $p < 0.01$); Cervical Lordosis and Kyphotic Angle (R = 0.63; $p < 0.001$); Coronal Imbalance and Spine Imbalance (R = −99; $p < 0.001$); Pelvic Obliquity and Pelvic Inclination (R = 0.87; $p < 0.001$).

Multiple linear regression models are shown in Table 2 and Figure 5. The adjusted R^2 value of the regression, that included Spine Length, Spine Inclination, and Kyphotic Angle as independent variables and Cervical Lordosis as the dependent variable, was 0.82.

Table 1. Descriptive statistics and unpaired *t*-test analysis between Cy-G and CG.

Spine Parameters	Cy-G (25)	CG (23)	df	*p*	Cohen's d (Effect Size)	95% Confidence Interval	
						Lower	Upper
Sagittal Plane							
Spine Length (mm)	450 ± 18.3	444 ± 45.8	22.0	0.621	0.1047	−0.3063	0.5134
Spine Inclination (°)	5.80 ± 3.50	3.25 ± 2.54	22.0	0.009	0.5981	0.1475	1.0372
Cervical Lordosis (mm)	58.2 ± 19.1	46.7 ± 20.4	22.0	0.106	0.3518	−0.0735	0.7696
Lumbar Lordosis (mm)	34 ± 12.4	48.3 ± 12.7	22.0	0.003	0.7110	−1.1635	−0.2456
Kyphotic Angle (°)	42.8 ± 10.0	42.7 ± 13.0	22.0	0.944	0.0148	−0.3941	0.4233
Lordotic Angle (°)	36.2 ± 10.3	42.3 ± 16.7	22.0	0.110	0.3472	−0.7647	0.0777
Frontal Plane							
Coronal Imbalance (mm)	3.04 ± 10.9	0.73 ± 7.68	22.0	0.371	0.1904	−0.2242	0.6008
Spine Imbalance (°)	−0.40 ± 1.38	−0.12 ± 0.986	22.0	0.413	0.1739	−0.5838	0.2399
Shoulder Obliquity (mm)	−1.36 ± 11.2	−1.17 ± 5.31	22.0	0.832	0.0449	−0.3645	0.4533
Shoulder Inclination (°)	−4.13 ± 19.4	−0.20 ± 1.04	22.0	0.345	0.2012	−0.6119	0.2140
Pelvic Obliquity (mm)	0.20 ±4.20	0.87 ± 2.93	22.0	0.735	0.0716	−0.4800	0.3384
Pelvic Inclination (°)	0.51 ± 2.73	0.55 ± 1.71	22.0	0.754	0.0662	−0.3437	0.4746

Table 2. Multiple linear regression models (dependent variable: Cervical Lordosis; independent variables: Spine Length, Spine Inclination, and Kyphotic Angle). Regression *p* value < 0.001; adjusted $R^2 = 0.82$.

Predictor	Estimate	SE	t	*p*
Intercept	−87.027	16.5516	−5.26	<0.001
Spine Length	0.190	0.0389	4.89	<0.001
Spine Inclination	3.458	0.3650	9.47	<0.001
Kyphotic Angle	0.887	0.1163	7.62	<0.001

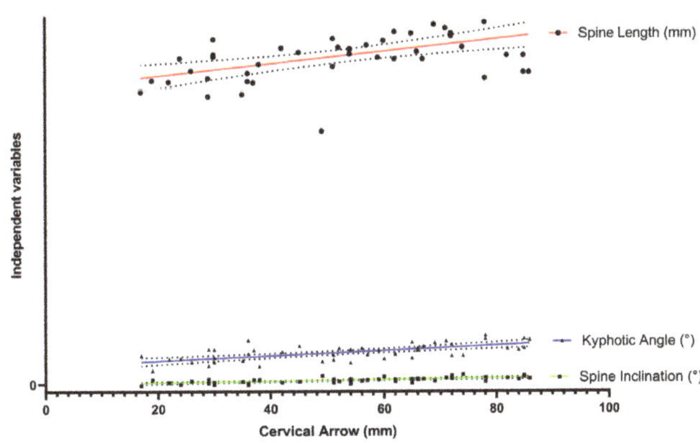

Figure 5. Relationship between cervical curvature and the independent variables: Spine Length, Kyphotic Angle, and Spine Inclination.

4. Discussion

The scientific literature lacks comprehensive information regarding spine morphology in standing position in recreational cyclists. This lack of knowledge is significant because the activation of spine and core muscles during prolonged periods of trunk flexion, such as during pedaling, can lead to stress on the spine and potentially contribute to LBP [44].

The aim of this study was to assess the spinal alignment in the upright posture among recreational cyclists and compare it with a control group of non-cyclist athletes. This evaluation was conducted using a cutting-edge 3D spinal imaging device. Our study is a case–control study, and upon reviewing the limited existing literature, we emphasize the necessity for additional studies with larger sample sizes.

Our findings detected two significant differences between the Cy-G and the CG, specifically in terms of the parameters of Spine Inclination and Lumbar Lordosis. Despite the literature suggesting that cycling might lead to an increase in thoracic kyphosis [9], our data indicate that this position does not yield significant differences in this part of the spine. More specifically, we observed that the Cy-G exhibited a decreased lumbar lordosis when compared to the CG. This observation suggests that cyclists' spines might undergo a postural adaptation, a conclusion in line with the existing literature on this topic [10,11,17].

Muyor et al. found a reduction in lumbar lordosis among cyclists [13]. This finding can be attributed to the tendency of cyclists to adopt a kyphotic posture in the lumbar region while pedaling [14]. Salai et al. showed a significant connection between saddle inclination and the incidence of LBP in cyclists [15]. As a matter of fact, the literature underscores that 41% of cyclists experiencing back pain require medical attention [45]. It would be advisable for athletes, athletic trainers, and coaches to pay attention to this result because a reduction in lumbar lordosis could increase the risk of spinal deficits, especially between L5 and the sacral plate due to considerable shear stresses [46]. The study by Rauter et al. is of particular interest because it showed that highly skilled road cyclists exhibit fewer body asymmetries compared to their less proficient counterparts [47]. This research demonstrated the efficiency of the 3D body scanning method in quickly and effectively identifying such asymmetries.

Moreover, our results showed a significantly greater spine inclination on the right side in the Cy-G compared to the CG. We speculate that this could be caused by an uneven distribution of forces exerted on the pedals, potentially leading to an unequal pressure distribution on the saddle and thereby accentuating lateral spinal inclination. These results are in agreement with previous studies that have also uncovered left–right asymmetries [47,48].

Mardsen et al. reported that LBP is a common overuse injury in cycling, and they showed the presence of a strength deficit of the pelvic stabilizer muscles in cyclists [49]. In a recent systematic review by Antequera-Vique et al., it was described that the practice of cycling produces adaptations in the morphology of the spine (i.e., lumbar flexion and a greater thoracic kyphosis) [9].

Although not statistically significant, we observed a greater cervical lordosis in the Cy-G compared to the CG, which is an interesting finding related to cyclists' riding posture. Our analysis revealed that factors like Spine Length, Spine Inclination, and Kyphotic Angle may serve as predictors for the increase in cervical lordosis. These findings could shed light on the prevalence of neck pain in cyclists, affecting up to 60% of them [50].

Furthermore, through Pearson's analysis, our data revealed potential relationships. These relationships were categorized as fair or good correlations. In a broader context, our data showed the impact of cervical lordosis on the kyphotic angle and the overall length of the spine. This holds significant relevance for cyclists, as previous studies have indicated a connection between cervical alignments and the compensatory mechanisms involved in maintaining a horizontal gaze [51,52]. Similarly, the analysis indicated an excellent relationship between the factors of Coronal Imbalance and Spine Imbalance, as well as between Pelvic Obliquity and Pelvic Inclination (respectively: $R = -0.99$; $p < 0.001$; $R = 0.87$; $p < 0.001$). This suggests that the assessment of the spine and the connection between upper and lower cervical alignment should be evaluated in terms of slope angles rather than relying solely on simple angles [51].

Our data imply that addressing the common issue of neck pain in cyclists requires a more comprehensive approach. It should go beyond mere symptom management or isolated muscle stretching programs for the cervical spine. Instead, a global strategy that encompasses the entire spine could be more appropriate. The literature indicates that integrating complementary postural training programs with specific training can effectively prevent LBP in cyclists [53,54]. Coaches should prioritize this integration in their programming. The primary strength of this study lies in its adequately sized sample of recreational cyclists, ensuring sufficient statistical power.

Limitations

The findings of this study are interesting and potentially valuable for future research. Nonetheless, some limitations need to be highlighted. First, our recruitment included cyclists of varying age groups, exclusively male. Second, while the literature assessed the instrument's reliability and found an excellent ICC, further studies with this instrumentation are necessary. Previous studies using the Spine3D system have been limited in terms of discussing their findings. To solidify our results, further studies are necessary. We hypothesize that factors contributing to our findings, apart from sitting position, may include imbalances in the forces exerted on the pedals [55].

5. Conclusions

In conclusion, our findings suggest that recreational cycling could lead to spinal deformities. This study is not aimed at discouraging the practice of recreational cycling due to the presence of these possible postural alterations. On the contrary, it is aimed at generating greater awareness of this sport in order to adopt specific training programs for the prevention of alterations in the musculoskeletal system [56]. Periodic evaluations could improve the methodology of training programs. Furthermore, the present study reveals the excellent applicability of Spine 3D and LiDAR technology in the sports field. This study aims to contribute to the knowledge on the adaptations of the spinal column of cyclists using rasterstereography. This technology has the potential to replace X-rays in the monitoring and preventing spinal deformities by contributing to reducing X-ray irradiation. Further studies are needed to confirm our hypotheses.

Author Contributions: A.B. and A.P. (Antonino Patti): Conceptualization; F.F., S.C., L.P. and G.B.: Data collection; E.T.: Data analysis; A.P. (Antonino Patti), E.T., V.G. and A.V.: Methodology; A.B. and A.P. (Antonio Palma): Supervision; A.A., G.M., I.R. and P.E.-G.: Visualization; A.P. (Antonino Patti): Writing—original draft; V.G.: Writing—review and editing. All authors have read and agreed to the published version of the manuscript.

Funding: The publication was made with the co-financing of the European Union—FESR or FSE, PON Research and Innovation 2014–2020—DM 1062/2021.

Institutional Review Board Statement: The study was carried out in compliance with the principles of the Declaration of Helsinki and approved by the Bioethics Committee of the University of Palermo (Num. 97/2022—Prot. 79743).

Informed Consent Statement: Informed consent was obtained from all subjects involved in the study.

Data Availability Statement: The data presented in this study are available on request from the corresponding author.

Conflicts of Interest: The authors declare no conflict of interest.

References

1. Fitzgibbon, S.; Vicenzino, B.; Sisto, S. Intervention at the foot-shoe-pedal interface in competitive cyclists. *Int. J. Sports Phys. Ther.* **2016**, *11*, 637–650. [PubMed]
2. Too, D. Biomechanics of cycling and factors affecting performance. *Sports Med.* **1990**, *10*, 286–302. [CrossRef] [PubMed]
3. Giustino, V.; Zangla, D.; Messina, G.; Pajaujiene, S.; Feka, K.; Battaglia, G.; Bianco, A.; Palma, A.; Patti, A. Kinematics of Cervical Spine during Rowing Ergometer at Different Stroke Rates in Young Rowers: A Pilot Study. *Int. J. Environ. Res. Public. Health* **2022**, *19*, 6790. [CrossRef] [PubMed]
4. Wojtys, E.; Ashton-Miller, J.; Huston, L.; Moga, P. The Association between Athletic Training Time and the Sagittal Curvature of the Immature Spine. *Am. J. Sports Med.* **2000**, *28*, 490–498. [CrossRef]
5. Chavarrias, M.; Carlos-Vivas, J.; Collado-Mateo, D.; Perez-Gomez, J. Health Benefits of Indoor Cycling: A Systematic Review. *Medicina* **2019**, *55*, 452. [CrossRef] [PubMed]

6. Bourne, J.E.; Sauchelli, S.; Perry, R.; Page, A.; Leary, S.; England, C.; Cooper, A.R. Health benefits of electrically-assisted cycling: A systematic review. *Int. J. Behav. Nutr. Phys. Act.* **2018**, *15*, 116. [CrossRef] [PubMed]
7. Oja, P.; Titze, S.; Bauman, A.; de Geus, B.; Krenn, P.; Reger-Nash, B.; Kohlberger, T. Health benefits of cycling: A systematic review. *Scand. J. Med. Sci. Sports* **2011**, *21*, 496–509. [CrossRef]
8. de Vey Mestdagh, K. Personal perspective: In search of an optimum cycling posture. *Appl. Ergon.* **1998**, *29*, 325–334. [CrossRef]
9. Antequera-Vique, J.A.; Oliva-Lozano, J.M.; Muyor, J.M. Effects of cycling on the morphology and spinal posture in professional and recreational cyclists: A systematic review. *Sports Biomech.* **2022**, *22*, 567–596. [CrossRef]
10. Beach, T.A.; Parkinson, R.J.; Stothart, J.P.; Callaghan, J.P. Effects of prolonged sitting on the passive flexion stiffness of the in vivo lumbar spine. *Spine J.* **2005**, *5*, 145–154. [CrossRef]
11. Muyor, J.M. The influence of handlebar-hands position on spinal posture in professional cyclists. *J. Back. Musculoskelet. Rehabil.* **2015**, *28*, 167–172. [CrossRef]
12. Muyor, J.M.; Zabala, M. Road Cycling and Mountain Biking Produces Adaptations on the Spine and Hamstring Extensibility. *Int. J. Sports Med.* **2016**, *37*, 43–49. [CrossRef] [PubMed]
13. Muyor, J.; Alacid, F.; López-Miñarro, P.; Casimiro, A. Evolution of Spinal Morphology and Pelvic Tilt in Cyclists of Different Ages. A Cross Sectional Study. *Int. J. Morphol.* **2012**, *30*, 199–204. [CrossRef]
14. Muyor, J.; López-Miñarro, P.; Alacid, F. Spinal Posture of Thoracic and Lumbar Spine and Pelvic Tilt in Highly Trained Cyclists. *J. Sports Sci. Med.* **2011**, *10*, 355–361. [PubMed]
15. Salai, M.; Brosh, T.; Blankstein, A.; Oran, A.; Chechik, A. Effect of changing the saddle angle on the incidence of low back pain in recreational bicyclists. *Br. J. Sports Med.* **1999**, *33*, 398–400. [CrossRef] [PubMed]
16. Bressel, E.; Larson, B.J. Bicycle seat designs and their effect on pelvic angle, trunk angle, and comfort. *Med. Sci. Sports Exerc.* **2003**, *35*, 327–332. [CrossRef] [PubMed]
17. Wilder, D.G.; Pope, M.H.; Frymoyer, J.W. The biomechanics of lumbar disc herniation and the effect of overload and instability. *J. Spinal Disord.* **1988**, *1*, 16–32. [CrossRef]
18. Nachemson, A. The load on lumbar disks in different positions of the body. *Clin. Orthop. Relat. Res.* **1966**, *45*, 107–122. [CrossRef]
19. Burnett, A.F.; Cornelius, M.W.; Dankaerts, W.; O'Sullivan, P.B. Spinal kinematics and trunk muscle activity in cyclists: A comparison between healthy controls and non-specific chronic low back pain subjects-a pilot investigation. *Man. Ther.* **2004**, *9*, 211–219. [CrossRef]
20. Taylor, D.C.; Dalton, J.D., Jr.; Seaber, A.V.; Garrett, W.E., Jr. Viscoelastic properties of muscle-tendon units. The biomechanical effects of stretching. *Am. J. Sports Med.* **1990**, *18*, 300–309. [CrossRef]
21. Wallace, A.L.; Hollinshead, R.M.; Frank, C.B. Creep behavior of a rabbit model of ligament laxity after electrothermal shrinkage in vivo. *Am. J. Sports Med.* **2002**, *30*, 98–102. [CrossRef] [PubMed]
22. Piotrowska, S.E.; Majchrzycki, M.; Rogala, P.; Mazurek-Sitarz, M. Lower extremity and spine pain in cyclists. *Ann. Agric. Environ. Med.* **2017**, *24*, 654–658. [CrossRef] [PubMed]
23. Cyr, A. Cervical Spine, Upper Extremity Neuropathies, and Overuse Injuries in Cyclists. *Phys. Med. Rehabil. Clin. N. Am.* **2022**, *33*, 187–199. [CrossRef] [PubMed]
24. Raso, V.J.; Lou, E.; Hill, D.L.; Mahood, J.K.; Moreau, M.J.; Durdle, N.G. Trunk distortion in adolescent idiopathic scoliosis. *J. Pediatr. Orthop.* **1998**, *18*, 222–226. [CrossRef]
25. Nash, C.L., Jr.; Gregg, E.C.; Brown, R.H.; Pillai, K. Risks of exposure to X-rays in patients undergoing long-term treatment for scoliosis. *J. Bone Jt. Surg. Am.* **1979**, *61*, 371–374. [CrossRef]
26. Betsch, M.; Wild, M.; Rath, B.; Tingart, M.; Schulze, A.; Quack, V. Radiation-free diagnosis of scoliosis: An overview of the surface and spine topography. *Orthopade* **2015**, *44*, 845–851. [CrossRef]
27. Betsch, M.; Wild, M.; Johnstone, B.; Jungbluth, P.; Hakimi, M.; Kuhlmann, B.; Rapp, W. Evaluation of a novel spine and surface topography system for dynamic spinal curvature analysis during gait. *PLoS ONE* **2013**, *8*, e70581. [CrossRef]
28. Willner, S. Moire topography for the diagnosis and documentation of scoliosis. *Acta Orthop. Scand.* **1979**, *50*, 295–302. [CrossRef]
29. Bugane, F.; Benedetti, M.G.; D'Angeli, V.; Leardini, A. Estimation of pelvis kinematics in level walking based on a single inertial sensor positioned close to the sacrum: Validation on healthy subjects with stereophotogrammetric system. *Biomed. Eng. Online* **2014**, *13*, 146. [CrossRef]
30. Bruni, D.; Bragonzoni, L.; Gagliardi, M.; Bontempi, M.; Akkawi, I.; Raspugli, G.F.; Iacono, F.; Patella, S.; Marcacci, M. Roentgen stereophotogrammetric analysis: An effective tool to predict implant survival after an all-poly unicompartmental knee arthroplasty-a 10 year follow-up study. *Knee Surg. Sports Traumatol. Arthrosc.* **2015**, *23*, 3273–3280. [CrossRef]
31. Marin, L.; Lovecchio, N.; Pedrotti, L.; Manzoni, F.; Febbi, M.; Albanese, I.; Patane, P.; Carnevale Pellino, V.; Vandoni, M. Acute Effects of Self-Correction on Spine Deviation and Balance in Adolescent Girls with Idiopathic Scoliosis. *Sensors* **2022**, *22*, 1883. [CrossRef]
32. Roggio, F.; Petrigna, L.; Trovato, B.; Zanghi, M.; Sortino, M.; Vitale, E.; Rapisarda, L.; Testa, G.; Pavone, V.; Pavone, P.; et al. Thermography and rasterstereography as a combined infrared method to assess the posture of healthy individuals. *Sci. Rep.* **2023**, *13*, 4263. [CrossRef]
33. Cuschieri, S. The STROBE guidelines. *Saudi J. Anaesth.* **2019**, *13*, S31–S34. [CrossRef]
34. Lucia, A.; Hoyos, J.; Chicharro, J.L. Physiology of professional road cycling. *Sports Med.* **2001**, *31*, 325–337. [CrossRef] [PubMed]

35. Bull, F.C.; Al-Ansari, S.S.; Biddle, S.; Borodulin, K.; Buman, M.P.; Cardon, G.; Carty, C.; Chaput, J.P.; Chastin, S.; Chou, R.; et al. World Health Organization 2020 guidelines on physical activity and sedentary behaviour. *Br. J. Sports Med.* **2020**, *54*, 1451–1462. [CrossRef]
36. Guidetti, L.; Bonavolonta, V.; Tito, A.; Reis, V.M.; Gallotta, M.C.; Baldari, C. Intra- and interday reliability of spine rasterstereography. *Biomed. Res. Int.* **2013**, *2013*, 745480. [CrossRef]
37. Rosner, B. *Fundamentals of Biostatistics*; Cengage Learning: Boston, MA, USA, 2015.
38. Patti, A.; Maggio, M.C.; Corsello, G.; Messina, G.; Iovane, A.; Palma, A. Evaluation of Fitness and the Balance Levels of Children with a Diagnosis of Juvenile Idiopathic Arthritis: A Pilot Study. *Int. J. Environ. Res. Public. Health* **2017**, *14*, 806. [CrossRef]
39. Kang, H. Sample size determination and power analysis using the G*Power software. *J. Educ. Eval. Health Prof.* **2021**, *18*, 17. [CrossRef] [PubMed]
40. Bowring, A.; Telschow, F.J.E.; Schwartzman, A.; Nichols, T.E. Confidence Sets for Cohen's d effect size images. *Neuroimage* **2021**, *226*, 117477. [CrossRef] [PubMed]
41. Sullivan, G.M.; Feinn, R. Using Effect Size-or Why the P Value Is Not Enough. *J. Grad. Med. Educ.* **2012**, *4*, 279–282. [CrossRef] [PubMed]
42. Koo, T.K.; Li, M.Y. A Guideline of Selecting and Reporting Intraclass Correlation Coefficients for Reliability Research. *J. Chiropr. Med.* **2016**, *15*, 155–163. [CrossRef] [PubMed]
43. Chicco, D.; Warrens, M.J.; Jurman, G. The coefficient of determination R-squared is more informative than SMAPE, MAE, MAPE, MSE and RMSE in regression analysis evaluation. *PeerJ Comput. Sci.* **2021**, *7*, e623. [CrossRef]
44. Streisfeld, G.M.; Bartoszek, C.; Creran, E.; Inge, B.; McShane, M.D.; Johnston, T. Relationship Between Body Positioning, Muscle Activity, and Spinal Kinematics in Cyclists With and Without Low Back Pain: A Systematic Review. *Sports Health* **2017**, *9*, 75–79. [CrossRef] [PubMed]
45. Clarsen, B.; Krosshaug, T.; Bahr, R. Overuse injuries in professional road cyclists. *Am. J. Sports Med.* **2010**, *38*, 2494–2501. [CrossRef]
46. Wang, Z.; Mac-Thiong, J.M.; Parent, S.; Petit, Y.; Labelle, H. The relevance of sacral and sacro-pelvic morphology in developmental lumbosacral spondylolisthesis: Are they equally important? *Eur. Spine J.* **2014**, *23*, 157–162. [CrossRef]
47. Rauter, S.; Simenko, J. Morphological Asymmetries Profile and the Difference between Low- and High-Performing Road Cyclists Using 3D Scanning. *Biology* **2021**, *10*, 1199. [CrossRef]
48. Rauter, S.; Vodicar, J.; Šimenko, J. Body Asymmetries in Young Male Road Cyclists. *Int. J. Morphol.* **2017**, *35*, 907–912. [CrossRef]
49. Marsden, M.; Schwellnus, M. Lower back pain in cyclists: A review of epidemiology, pathomechanics and risk factors. *Int. Sportmed. J.* **2010**, *11*, 216–225.
50. Mellion, M.B. Common cycling injuries. Management and prevention. *Sports Med.* **1991**, *11*, 52–70. [CrossRef]
51. Iplikcioglu, A.C.; Karabag, H. Analysis of Components of Upper Cervical Lordosis in Asymptomatic Lordotic and Kyphotic Subjects. *World Neurosurg.* **2023**, *171*, e852–e858. [CrossRef] [PubMed]
52. Tan, L.A.; Riew, K.D.; Traynelis, V.C. Cervical Spine Deformity-Part 1: Biomechanics, Radiographic Parameters, and Classification. *Neurosurgery* **2017**, *81*, 197–203. [CrossRef]
53. Greco, G.; Messina, G.; Angiulli, A.; Patti, A.; Iovane, A.; Fischetti, F. A preliminary comparative study on the effects of Pilates training on physical fitness of young female volleyball players. *Acta Medica Mediterr.* **2019**, *2*, 783. [CrossRef]
54. Patti, A.; Thornton, J.S.; Giustino, V.; Drid, P.; Paoli, A.; Schulz, J.M.; Palma, A.; Bianco, A. Effectiveness of Pilates exercise on low back pain: A systematic review with meta-analysis. *Disabil. Rehabil.* **2023**, 1–14. [CrossRef] [PubMed]
55. Vicari, D.S.S.; Patti, A.; Giustino, V.; Figlioli, F.; Alamia, G.; Palma, A.; Bianco, A. Saddle Pressures Factors in Road and Off-Road Cyclists of Both Genders: A Narrative Review. *J. Funct. Morphol. Kinesiol.* **2023**, *8*, 71. [CrossRef] [PubMed]
56. Giustino, V.; Messina, G.; Patti, A.; Padua, E.; Zangla, D.; Drid, P.; Battaglia, G.; Palma, A.; Bianco, A. Effects of a Postural Exercise Program on Vertical Jump Height in Young Female Volleyball Players with Knee Valgus. *Int. J. Environ. Res. Public. Health* **2022**, *19*, 3953. [CrossRef]

sports

Article

Artificial Intelligence Program for Predicting Wrestlers' Sports Performances

Roman Sergeevich Nagovitsyn [1,2,*], Roza Alexeevna Valeeva [3] and Liliia Agzamovna Latypova [4]

1 Faculty of Pedagogical and Art Education, Glazov State Pedagogical Institute, 427621 Glazov, Russia
2 Department of Methodology and Technology of Universal Competencies, Kazan State Institute of Culture, 420059 Kazan, Russia
3 Institute of Psychology and Education, Kazan Federal University, 420008 Kazan, Russia; valeykin@yandex.ru
4 Institute of Management, Economics and Finance, Kazan Federal University, 420008 Kazan, Russia; melilek@yandex.ru
* Correspondence: gto18@mail.ru

Abstract: To date, there are conflicting opinions about the effectiveness of the introduction of artificial intelligence technologies in sports. In this regard, the purpose of the study was to develop and integrate an intellectual program for predicting competitive success into the process of selecting wrestlers to increase its effectiveness. The authors developed a program for predicting the sports performance of wrestlers on the basis of artificial intelligence technology. To implement the study, the individual data of Greco-Roman wrestlers (n = 72) were collected and processed on 36 comparison traits, ranked into categories according to three key areas: sports space, hereditary data and individual achievements. As a result of data processing through means of deep neural networks and machine learning algorithms, two prediction categories were identified: athletes who performed at the sport rank or the highest standard and athletes who did not achieve this standard. Control testing of the created program showed only 11% of error probability in predicting a given wrestler's competitive performance. As for the functionality of the program in the area of classification of the features by category, the authors' artificial intelligence program with 100% probability identified key categories of traits that reliably affect the results of the future sports performance of a young wrestler. Thus, the use of neural networks and machine learning algorithms, according to the results of the study, improves the quality of sports selection, which will allow further timely individualization and improvement of the training process of young wrestlers.

Keywords: sports performance; artificial intelligence; wrestlers; prediction; program

check for updates

Citation: Nagovitsyn, R.S.; Valeeva, R.A.; Latypova, L.A. Artificial Intelligence Program for Predicting Wrestlers' Sports Performances. *Sports* **2023**, *11*, 196. https://doi.org/10.3390/sports11100196

Academic Editors: Rodrigo Zacca, Flávio Antônio de Souza Castro and Rui Miguel Simões de Azevedo

Received: 28 July 2023
Revised: 6 September 2023
Accepted: 4 October 2023
Published: 8 October 2023

1. Introduction

Information technology has penetrated into the everyday life of young people, students and society as a whole [1]. Every year, the practical implementation of the achievements of scientific and technological progress in the field of digitalization begins to more significantly affect any of the spheres of human social activity [2]. The introduction of various information programs, neural networks and artificial intelligence (AI) technologies in all spheres of society has reached a stage where the general social system can no longer fully exist without this integration [3]. The implementation of information innovations, including the most relevant of them such as AI programs and machine learning (ML) programs, significantly increase the effectiveness of continuous the social modernization conditions for the quality of human life [4].

Big data, ML and AI, providing the faster development of technology toward human-centric applications, are beginning to globally and fundamentally change the way society and technology develop [5]. With the advent of the Internet of Things, through the use of data-driven conceptual approaches, there are more opportunities to activate new channels for the dissemination of innovations locally and globally in the economy, politics, medicine, education and in all areas of human life [5]. Based on AI technologies, it has become possible to universally implement smart and lean manufacturing, as an integral part of Industry 4.0., in six main aspects of the production line: quality checks, maintenance, reliable design, environmental impact reduction, factory integration and after-production support [6]. Various intellectual platforms are being systematically created through the processes of modeling and marketing simulation to create an analytical ecosystem as one of the main tools for positive social dynamics [7]. Based on the instant processing of big data and their further use, great prospects open up for stimulating innovation and improving public administration, which leads to an increase in the quality of life of the population [7].

As research shows, for many centuries, data mining has been used in various spheres of human activity to process the results of the observations and measurements of real objects, as well as in natural and computational experiments [8,9]. Thus, there is an interesting study on the creation of effective medical services through big data analysis, which proves that multimodal intelligent forecasting can reliably assess the applicability and effectiveness of healthcare policies [10]. On the basis of AI, practical variations of medical services for disease prevention and integrated health policy are ranked through a combination with specific segments of the population, taking into account social problems and spatial and temporal properties [10]. With the help of intellectual technologies, a detailed psychological analysis of college students is implemented when performing various types of physical activity [8]. To date, it has become possible to analyze the indicators of physical health and educational characteristics of students at various stages of sports training based on AI [9]. In the system of physical education and sport as one of the key spheres of society, which has recently been supported by information technologies—and in some cases even artificial intelligence—it is crucial to realize the role of these technologies in increasing the effectiveness of the training process [11,12]. The active use of these innovative technologies increases the reliability of the collection of system data, their accuracy and interpretability for its subsequent analysis to develop an information management system for assessing the physical fitness of athletes [11]. The use of intelligent algorithms improves the quality of refereeing in sports matches and other sports, through an increase in the percentage of objectivity, accuracy, speed of scoring and determining results, as well as timely detection of violations of the rules of athletes [12]. As program and regulatory documents show, the active, practical implementation of intelligent technologies contributes to the creation of effective conditions for improving the quality of life of the population through the intellectualization of services in the sphere of physical education, fitness and sport [13,14]. Additionally, it promotes efficiency of the system analysis of training indicators for professional orientation optimization and early detection of children with outstanding abilities, automation of knowledge quality assessment and analysis of information on the results of the training process [4,15].

In this direction, it is AI that is positioned as a worthy innovative tool able to provide large-scale personalized physical education and training in sports based on big data [16,17]. With the help of artificial neural networks, based on the identification of key performance indicators, it is possible to determine playing talents in professional football as well as their status as field players in the football league [16]. Models are being developed for predicting the quantitative values of physiological indicators of the work of an athlete's body at the level of anaerobic metabolism threshold when performing ergo-spirometric load testing to failure using ML algorithms [17]. The use of AI helps in the field of sports medicine, namely in the creation of predictive models in the healthcare industry to assess the risk and predict the occurrence of various sports injuries, maintain health stability and physical performance through the analysis of data on internal and external factors [18]. To analyze the technical training of combatants, new intelligent optimization algorithms are being actively used, based on evolutionary calculations through the identification of biomechanical dependencies in martial arts data and the complex neuromuscular control of athletes [19]. Some current empirical studies have proven that the use of intelligent technologies can partially replace coaches and teachers in various organizations of sports profiles and improve the quality of physical education and sports training [20,21]. AI programs help track players, evaluate the effectiveness of each tactical action, create the optimal ratio of sports risk and potential gain in confrontation by calculating useful indicators without the use of painstaking human labeling [20]. Based on the identification and comparison of physiological and positional data during a sports game in the context of a real match, machine learning technologies reliably recognize subsequent technical actions, which makes it possible to correct the managerial decisions of the coaching staff and increase the team's performance [21].

In the specialized literature, the use of AI technologies on the analysis of big data in the field of sports activities is justified as one of the significant and effective tools to improve the quality of health and motor performance of novice and professional athletes [3,8,10]. It offers solutions to many problems, such as the differentiation of engaged persons to identify dependencies and relationships between them, analysis of the quality of sports training and the risk of trainees receiving poor physical development, automatic construction of recommendations on the use of information resources and materials to more effectively master the sports program [15,22,23].

In the last decade, innovative intelligent technologies have significantly increased their impact on the field of sports, fitness and physical education [16,24]. Intelligent technologies are being used to analyze big sports data, in the aspect of statistics of athletes' and referees' actions from match videos or physical activity metrics of competitors [4,14,25]. Based on the use of special sensors and the processing of the received data via machine learning with the help of the neural networks of a deep closed recurrent unit, medical and biological monitoring of the training process is carried out by comparing the electrocardiogram data of athletes and stress, represented by the features of their heart rate variability, which further allow us to model features of psychological stress involved in various sports conditions and implement preventive procedures [26]. AI technologies provide virtual advisory assistance, effectively promoting in the support of athletes' physical and sports potential and in the realization of cyber sports and sports betting [15,20]. In the pre-competitive stage of sports training, athletes, together with the coaching staff, begin to actively use the results of intellectual analysis to improve the effectiveness of preparation for performance at competitions, as well as AI technologies that allow fully automating the athlete's pre-start training [10,14].

Indeed, there are conflicting views in the specialized literature on the effectiveness of AI in sports [1,24]; the introduction of these technologies without the implementation of reliable experimental studies may be negative, with the consequence that the next generation will be ill-prepared for a dynamic and changing world with the increasing development of big data [21,23]. However, the purpose of this research is not to introduce global standards based on the introduction of intelligent technologies and certainly not to find ways to replace trainers [4]. The authors aim to create an AI program that, when implemented, will enable coaches and the administration of sports organizations to reliably improve the effectiveness of sports training based on big data [8].

Thus, a preliminary overview analysis of the existing possibilities for the implementation of AI technologies to improve the efficiency of sports training of athletes shows that at the moment there is a special need for the implementation of such an experiment. In this regard, the authors set the goal of the study: to experimentally implement a program for predicting the sports results of athletes in the context of the practical implementation of AI technologies.

Taking into account the issues discussed above, the hypothesis of the study is the following—the implementation of sports selection for sports performance in wrestling will be more effective if an intelligent program for predicting competitive success is developed and implemented to help in the individualization of the training process.

This program would be based on the use of various ML algorithms and big data analysis. Its practical implementation in a sports organization will make it possible to implement the recommendation to identify successful young athletes who have certain types of predispositions to obtain the highest results. The implementation of this program will make it possible to make recommendations for the sports career of a young athlete at the stage of further choosing his sports trajectory, then again at the stage of his training in sports and recreation groups [23]. In this regard, the use of AI in the analysis of big data will help to correct the sports trajectory of a novice athlete [15] and, in the future, will also provide an opportunity for administrative and coaching staff to adjust and personalize sports training at the stage of sports improvement.

2. Materials and Methods

During the experimental study (February–September, 2022), the collection and processing of personal data of the wrestlers (n = 72) engaged in Greco-Roman wrestling or who have already completed their wrestling career were carried out. To collect the information, the authors analyzed the archived data (1988–2008) of the wrestlers from the cities of Glazov (the Udmurt Republic, Russia) and Kazan (the Republic of Tatarstan, Russia) and conducted surveys on social networks and via phone. For the study, they collected data from the athletes in the time period when they were just starting their sports career and were young wrestlers. All data were ranked in three directions ("Hereditary Data", "Sports Space" and "Individual Achievements") and were collected in a common dataset. Figure 1 below shows a snapshot of the collected datasets of 12 wrestlers for "Sports Space" and "Individual Achievements".

N/Trait	SS1	SS2	SS3	SS4	SS5	SS6	SS7	SS8	SS9	SS10	SS11	SS12	IA1	IA2	IA3	IA4	IA5	IA6	IA7	IA8	IA9	IA10	IA11
1	0	2	1	2	1	0	0	2	1	0	0	2	1	1	2	2	0	1	2	0	2	0	0
2	0	1	2	0	1	0	0	1	1	0	1	1	2	2	1	2	0	0	0	2	1	1	0
3	1	2	1	2	0	1	1	2	2	0	2	3	1	2	2	1	0	0	2	1	0	0	0
4	1	1	2	1	0	0	2	1	1	1	1	2	1	2	2	0	2	1	1	0	2	1	1
5	0	2	1	0	1	0	2	1	1	0	1	1	2	2	2	2	1	2	1	1	0	2	2
6	2	2	1	2	0	0	0	1	3	0	1	1	2	1	2	1	2	1	1	0	1	2	1
7	0	2	1	1	1	1	0	1	2	2	0	1	2	1	1	0	0	1	0	0	1	1	2
8	1	2	2	2	1	2	1	0	1	1	0	1	0	1	1	0	2	1	2	1	1	0	2
9	1	0	1	0	2	0	1	1	1	0	0	0	2	0	0	0	0	0	1	1	1	0	1
10	2	2	1	0	1	0	1	1	3	1	0	0	2	1	0	0	0	0	0	1	2	1	2
11	1	1	2	1	2	1	2	1	3	0	0	1	2	2	0	1	1	1	0	2	1	2	0
12	0	1	2	1	2	2	1	2	2	2	2	1	1	1	2	2	1	2	1	1	0	1	0

Figure 1. Snapshot of collected datasets of 12 wrestlers for "Sports Space" and "Individual Achievements".

The focus group for the experiment included wrestlers who currently have the highest sports category "Candidate for Master of Sports of Russia" or the sports title "Master of Sports of Russia" (n = 19), athletes in III–I sports categories (n = 21) and wrestlers who have junior sports categories or have no categories (n = 32). In the experiment, the data of only those athletes (aged 27 to 47 years today) who voluntarily agreed to take part in the experiment and provided the necessary archival information were analyzed.

The study included the development of an intellectual program for creating forecasts of the competitive success of athletes using the Orange analytics system. A workflow for data analysis through means of process intelligence models was created in the interface of the Orange software. Classical deep networks and special machine learning algorithms for categorical classification, "Logistic Regression" and "Random Forest", were used to implement the experimental workflow [4]. The analytical system Orange was chosen for the study—as this is an open source program for ML, statistical research and data visualization—which has a large set of research and analytical functions. Compared with Megaputer PolyAnalyst, TIBCO Data Science, Polymatica and Anaconda analog intelligent platforms, the Orange Data Mining software uses visual programming, which is implemented via a user-friendly graphical interface with a variety of free analytical functionality and classification algorithms for convenient operation by users without special technical education.

3. Results

The predictive and classification functionality of the Orange intelligent platform was used in the analysis of the big data of wrestlers from the focus group. To "train" the author's program on the Orange smart platform, not all wrestlers' data were used but only randomly selected data (n = 36) for pre-testing.

As a result of the study, the system analysis of the program created at the previous stage to predict the competitive performance of young wrestlers in the intelligent system Orange allowed the determination of the basic classification for the prediction of sports performance. The program was finally tested with a validation sample (n = 18) selected randomly but with the mandatory condition of sampling uniformity. Using the data from the validation sample, the program independently identified the prognosis on the basis of special intelligent algorithms and neural networks, taking into account the classification of regularities. When an error was detected, additional "training" of the intelligent system was performed until the moment when the created program could independently find the optimal numerical patterns to predict the correct sports result [4]. In conclusion, the effectiveness of the created intellectual program was finally tested using the data of the wrestlers of the control sample (n = 18), which was randomly formed, but on the condition that it included wrestlers with titles, categories, and without sports wrestling qualifications. The data of the wrestlers of the control sample, analyzed with the intellectual program created in the experiment, made it possible to identify the resulting percentage of reliability in the correctness of the forecast proposed by the author's development under the conditions of using AI.

On the basis of the preliminary analysis of the specialized scientific literature and the peculiarities of young athletes' versatile monitoring, a system of traits and their categories for the intelligent prediction was developed [1,4,9,12,24,25]. The system was classified into 36 traits, each of which was ranked into 2–4 categories, in three main personal areas: 13 traits via hereditary data, 12 traits via sports space, and 11 traits via individual achievement (Table 1).

Table 1. Classification of young wrestlers' data via traits and their categories.

Designation	Trait	Categories by Code Numbers
		Hereditary Data
HD1	Sex	Young men (1), young lady (0)
HD2	Zodiac sign by the elements	Fire (0), earth (1), air (2), water (3)
HD3	Decade of the zodiac sign	20–28/29/30/31 (0), 1–10 (1), 11–19 (2)
HD4	Maximum oxygen intake	<25 (0), 25–35 (1), >35 (2)
HD5	Blood group	1 (0), 2 (1), 3 (2), 4 (3)
HD6	Hemoglobin	<115 (0), 115–150 (1), >150 (2)
HD7	Hematocrit	<35 (0), 35–45 (1), >45 (2)
HD8	Minerals (%)	<7 (0), 7–10 (1), >10 (2)

Table 1. *Cont.*

Designation	Trait	Categories by Code Numbers
HD9	Total fluid (%)	<40 (0), 40–50 (1), >50 (2)
HD10	Protein (%)	<10 (0), 10–15 (1), >15 (2)
HD11	Muscle fibers white/red	<w (0), w = r (1), <r (2)
HD12	Lung vital capacity (mL)	<1500 (0), 1500–2000 (1), >2000 (2)
HD13	Respiratory volume (mL)	<180 (0), 180–250 (1), >250 (2)
		Sports Space
SS1	Age at the time of "enrollment" in the sports section (year)	7–8 (0), 9–10 (1), 11–12 (2)
SS2	Family status	Orphan (0), single–parent family (1), nuclear family (2)
SS3	Place of residence	Village (0), town with the population of less than 100 thousand residents (1), town with the population of more than 100 thousand residents (2)
SS4	Siblings (elder brothers or sisters)	No siblings (father/mother) or does not practice any sports (0), has an athletic category (1), has a sports title (2)
SS5	Father	
SS6	Mother	
SS7	Coach qualifications	Has an athletic category (0), has the title of the Candidate for Master of Sports (CMS) or the athletic title of the Master of Sports (MS) (1), has the title of the International Master of Sports (IMS) or the Merited Master of Sports (2)
SS8	Results of the coach trainees over the period of the last 5 years	Athletic categories (0), athletic title of CMS (1), athletic title of MS or IMS (2)
SS9	Coach experience	<5 years (0), 5–15 years (1), 16–30 years (2), >30 years (3)
SS10	Father or mother works in physical education and sports	No (0), yes (1), part-time (2)
SS11	Type of educational institution	School (0), gymnasium (1), lyceum (2)
SS12	Who signed up for the sports section	Him- or herself or a friend (0), grandparent (1), brother or sister (2), father or mother (3)
		Individual Achievements
IA1	Average score at school for the previous year	<4.2 (0), 4.2–4.7 (1), >4.7 (2)
IA2	Ratio of trainings missed to the number of all classes (%)	>10 (0), 5–10 (1), <5 (2)
IA3	Physical development in accordance with the "Ready for Labor and Defense" test	No pin badge (0), bronze/silver pin badge (1), gold pin badge (2)
IA4	Average performance in the first two or three competitions (place)	<3 (0), 2–3 (1), >2 (2)
IA5	Cooper's strength test (min)	<1.2 (0), 1.2–2 (1), >2 (2)
IA6	Muscle mass index to height and weight	<15 (0), 15–20 (1), >20 (2)
IA7	Fat content (%)	<25 (0), 25–35 (1), >35 (2)
IA8	Fat-free mass (%)	<75 (0), 75-65 (1), >65 (2)
IA9	Muscle mass in relation of the lower limbs to the upper limbs	<1.2 (0), 1.2–1.5 (1), >1.5 (2)

Table 1. *Cont.*

Designation	Trait	Categories by Code Numbers
IA10	Muscle mass in relation to right and left extremities	<1.05 (0), 1.05–1.1 (1), >1.1 (2)
IA11	Muscle mass in relation of limbs to body	<1,2 (0), 1.3–1.5 (1), >1.5 (2)

Designation—number encoding of each of the characteristics of the ranking of these wrestlers; Trait—the main directions of wrestlers' characteristics in the three areas of "Hereditary Data", "Sports Space" and "Individual Achievements"; Categories by code numbers—categories and indicators of traits in coding for data processing in the Orange system.

After determining the comparative traits by category, the personal data of Greco-Roman wrestlers participating in the experiment, obtained at a time when they were just starting their sports career and participated in their first competitions, were loaded onto the Orange analytical system. As a result of a comparative analysis of the received data of wrestlers based on AI technology, two main levels of sports success prediction were eventually ranked: Greco-Roman wrestlers who had fulfilled the sport rank or achieved the highest level and the wrestlers who had not reached this level. After additional "training" of the intellectual system with validation sampling, the program was tested with data from a control sample (n = 18). As a result of intellectual data processing, only two errors out of the 18 tested data points of wrestlers of the control test sample were reliably recorded, which ultimately indicates an 11% probability of an incorrect answer by the author's program in the implementation of forecasts of the competitive success of wrestlers. And it should also be specially noted that during the implementation of the intellectual program, only one error was detected out of the 10 athletes when testing the control sample for a group of 10 athletes who did not achieve a high sports result in Greco-Roman wrestling.

Thus, the author's program for predicting the competitive performance of athletes based on intellectual technologies has been experimentally tested and is ready for practical implementation. Nevertheless, during its practical implementation in the process of sports training of Greco-Roman wrestlers, additional data were required on which traits and categories had a greater impact on performance or, conversely, on the irrelevance for further success in competitive activity. Accordingly, after alternating the features and functionality of the intelligent program, we managed to determine the trait categories most reliably influencing the result "Fulfilling the highest standards for the Candidate for Master of Sports of Russia or the Master of Sports of Russia" (Figure 2).

Figure 2 shows the process of classification of traits and categories in the implementation of the authors' developed program on the Orange platform. On the left side of Figure 2, the introduction of 36 traits is marked by categorical classification + 1 trait, as an achieved result (will or will not fulfill). This field also denotes the experimental samples (n = 72) that were analyzed. The middle part of Figure 2 shows the workflow of predicting and classifying the loaded data on the intelligent platform by selecting widgets and connections between them. The right margin shows the result of traits classification by category, ranked above in Table 1. As it can be seen from Figure 2, the program determined the key values according to the result "Fulfilling the highest standards for the Candidate or Master of Sports of Russia" in the following sequence in descending order: SS6(2), SS4(2), SS8(2), SS9(2), SS10(2), SS3(2), IA3(2), ... The detected categories of traits indicate their most reliable significance in predicting the wrestlers' competitive performances.

In turn, with further changing of the intelligent program target setting, there were identified categories of traits that most reliably influence the result "Not fulfilling the highest standards of Candidate for Master of Sports of Russia or the title of Master of Sports of Russia" (Figure 3).

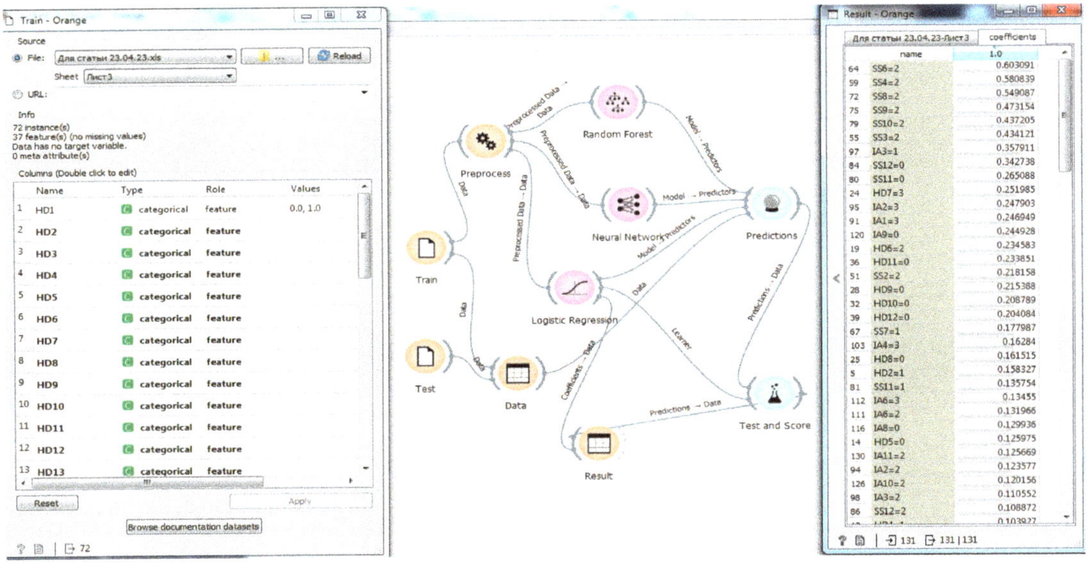

Figure 2. The working window of the developed program on the Orange platform to classify the traits according to the result "Fulfilling the highest standards of the Candidate for Master of Sports of Russia or the Master of Sports of Russia".

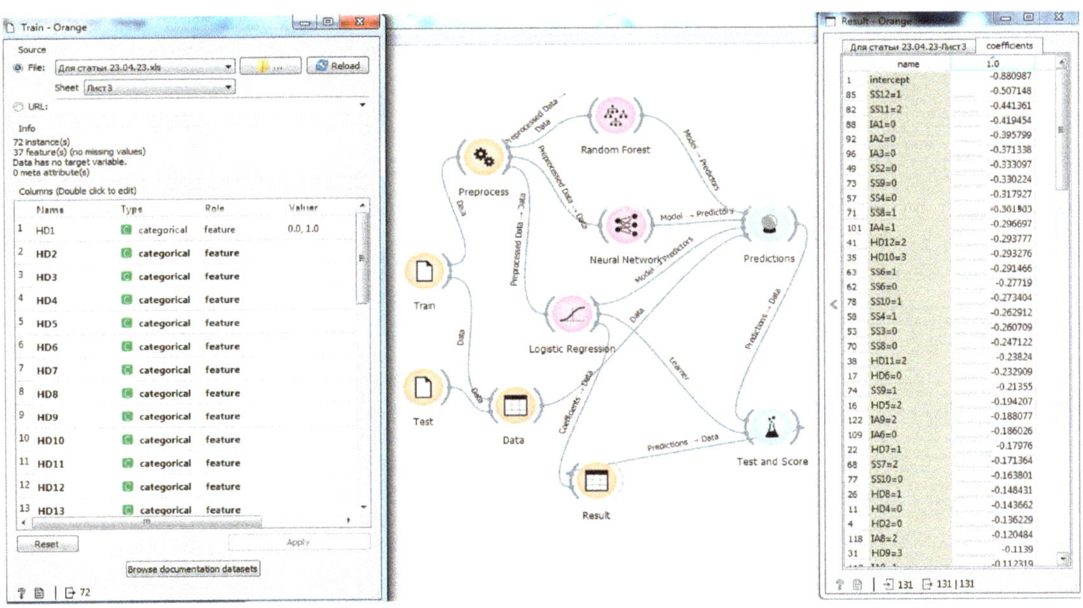

Figure 3. The working window of the developed program on the Orange platform to classify the traits according to the result "Not fulfilling the highest standards of Candidate for Master of Sports of Russia or the title of Master of Sports of Russia".

Figure 3 shows the same process of the classification of traits and categories in the implementation of the authors' program on the Orange platform. Figure 3 demonstrates that the program determined the key values for the opposite result of "Not fulfilling the highest standards of Candidate for Master of Sports of Russia or the title of Master of Sports of Russia" in the following sequence in descending order: SS12(1), SS11(2), IA1(0), IA2(0), IA3(0), SS2(0), SS9(0), ... The identified trait categories indicate their most reliable significance in predicting the competitive performance of wrestlers by the designated negative result for the given study.

After identifying the trait categories that most reliably influence the result of predicting the competitive performance of young wrestlers by analyzing the data from the entire experimental sample (six categories for the positive and six for the negative), the percentage ratio in different combinations of trait categories was obtained (Table 2).

Table 2. Results of reliability in % of categories of traits, credibly affecting the prediction of competitive performance of young wrestlers.

Categories of traits on the result "Fulfilling the highest standards of Candidate Master of Sports of Russia or the title of Master of Sports of Russia"		1 *	1–2 **	1–3 ***
1	Athletic mother: has an athletic rank **SS6(2)**/ elder brother or sister: has sports rank **SS4(2)**	>85		
2	Efficiency of trainees in the last 5 years: title of Master of Sports /International Master of Sports **SS8(2)**/trainer's work experience: from 16 to 30 years **SS9(2)**		>92	
3	Father or mother works in the field of physical education and sports: part-time **SS10(2)**/place of residence: city of more than 100,000 citizens **SS3(2)**			100
Categories of traits according to the result "Not fulfilling the highest standards of Candidate Master of Sports of Russia or the title of Master of Sports of Russia"		1	1–2	1–3
1	Who "enrolled" a young athlete in the sports section: grandmother or grandfather **SS12(1)**/ training of a young athlete: lyceum **SS11(2)**	>82		
2	Previous year's high school grade point average: <4.2 **IA1(0)**/training absences to the number of all classes (%): >10 **IA2(0)**		>89	
3	Physical development on the "Ready for Labor and Defense" test: no pin badge **IA3(0)**/young athlete's family status: orphan **SS2(0)**			100

* If a wrestler of the first category of traits (one of the two options) is revealed, the corresponding percentage result of the reliability of the sports result is recorded; ** if a wrestler has the first and second categories of traits (one of the two options), the corresponding percentage result of the reliability of the sports result is recorded; *** if a wrestler has three categories of traits (one of the two options), the corresponding percentage result of the reliability of the sports result is recorded.

Prediction efficiency in both categories is 100% fixed if in monitoring a young athlete shows three categories of traits together on one of the different options out of every two for each trait. Thus, the program has 100% probability of detecting a reliable prediction when finding three traits in a young wrestler.

In turn, if a young wrestler's mother is engaged in sports and has an athletic title or one of his elder brothers or sisters has a sports title (one trait with a category in the performance), there is more than 85% probability that he will be successful in future competitive activities in wrestling. And if he is also being coached by a mentor with work experience from 16 to 30 years or if the results of his previous trainees in the last 5 years are not less than Master of Sports of Russia or Master of Sports of International class (categories 1 and 2), then the probability of achieving the highest sports category and Russian rank increases up to 92%.

On the other hand, if a young wrestler was "signed up" by his grandmother or grandfather or was studying in a sports section in a non-sports lyceum (the first indication with the category of not achieving sports results), then he has more than 82% probability to not achieve a high sports level in his career in the future. And if he misses more than 10% of training sessions or his grade point average at school for the previous year before he was engaged in the wrestling section is less than 4.2, then the probability of not achieving the highest sports category and the title of Russia increases to 89%.

4. Discussion

The problem of sports selection at the early stages of young athletes' sports training, as the results of the present study showed, implies the search for regularities, and in the given case, the highest combination of traits by categories compared to the combination of indicators of the whole studied group [4]. It, in turn, comes to the task of ordering system features comparing the indicators of young wrestlers and output results in the form of predictive models based on big data analysis [15,25]. This technology is based on determining the so-called taxonomic distance, that is, the distance between points in the multidimensional space from the identified traits via hereditary data, sport space and individual achievements of young wrestlers [3,10]. As the results of the study demonstrated, the creation of an intelligent system of patterns using comparison traits reached 100%, when at least three traits in the categories coincided. And at the coincidence of only one category of a trait, the probability became 85% on achieving the performance of the highest category of the Candidate for the Master of Sports of Russia or a rank of the Master of Sports of Russia and 82% on not achieving the performance of the highest category or a sports rank.

In the modern period of sports and physical education development, many mathematical methods are used to predict future sports performance at different stages of sports improvement, which is associated with a number of technical and energy-consuming problems in analyzing big data [3,8,15]. The low level of reliability in identifying a meaningful sports selection strategy has a number of important features in creating predictions [16,21], namely, insufficient identification of individual hereditary characteristics and achievements of young athletes, identification and adjustment of traits and categories included in the development, lack of timely risk prediction and limitations on practical applicability [23,25]. In turn, the implementation of AI programs can increase the reliability of prediction detection and significantly improve the accuracy in the processing of athletes' data when entering the sports section [7,20]. The use of various ML algorithms and neural networks will make it possible to more accurately identify trainees who are not predisposed to a particular sport, which will further open the trajectory of educational impact and individualization of the training process for this group of athletes.

Despite the fact that the use of various ML algorithms and neural networks makes it possible to implement recommendations for identifying athletes who are not predisposed to a wrestling career, this study has a number of significant limitations. In the study, more emphasis was placed on the sections of data on athletes' "Sports Space" and "Individual Achievements" and less in the direction of "Hereditary Data", which could significantly affect the effectiveness of the forecast. In the experiment, only the data of athletes in an individual sport, in particular in Greco-Roman wrestling, were taken, which also has significant limitations for using the recommendations of the program in other sports, particularly in team sports. Nevertheless, the methodological recommendations proposed by the developed program can further open a trajectory for the educational impact and individualization of the training process for this group of athletes.

5. Conclusions

Thus, in the presented experiment, an attempt was made to look into the future and introduce the intellectual technology of big data into the process of sports training of wrestlers, which is fully consistent with the development of modern martial arts. The developed authors' program on the basis of AI allows the prediction of the competitive performance of a novice wrestler and reliably captures the main significant categories of traits that positively or negatively affect the prospect of a novice Greco-Roman wrestler achieving the highest sports grade or title of Russia in the future. The practical significance of the study on the implementation of neural networks and ML algorithms was confirmed with the reliable experimental data. Accordingly, the author's development will improve the quality of the sports selection of young wrestlers and also, in a recommendatory aspect, contribute to the timely personalization and adjustment of the training process of young athletes. In conclusion, it is the artificial intelligence technology with properly selected algorithms for the processes of intellectual clustering, classification and forecasting that has a strong and stable ability for reliable sports analysis, effective selection of the main criteria and indicators that are significant for improving the efficiency of a young wrestler's sports trajectory.

The results of the study will be of interest to coaches of children's youth sports schools and schools of higher sportsmanship in martial arts. Practical recommendations for loading and processing wrestlers' data into the program developed in the study will allow methodologists of sports organizations to timely implement the sports selection of beginner athletes and systematically adjust the training process at the early stages of wrestlers' training. The peculiarities of the study by stages and the results and conclusions obtained during the implementation of the experiment were repeatedly discussed at methodological meetings and round tables with representatives of sports organizations in the field of Greco-Roman wrestling. In the future, information about the results of this experiment will be disseminated as part of a special module at advanced training courses for wrestling coaches in youth sports schools.

The next study of the authors in the field of the implementation of AI in sports will continue the experiment to supplement big data from other types of martial arts: freestyle wrestling, judo and taekwondo. The experiment will continue to correct the program using additional intelligent algorithms, new systems of neural networks, implementation of a special n-fold cross-validation for higher reliability of the results and the introduction of new features for classifying the processed data. In the future, the main focus of data collection will be focused on the search for the biomedical, psychological and physiological data of young athletes in order to create more accurate predictions of their martial arts sports career.

Author Contributions: Conceptualization: R.A.V. and R.S.N.; methodology, R.S.N.; software, R.S.N.; validation, R.S.N., R.A.V. and L.A.L.; formal analysis, L.A.L.; investigation, L.A.L.; resources, R.S.N.; data curation, R.S.N.; writing—original draft preparation, R.A.V.; writing—review and editing, L.A.L.; visualization, L.A.L.; supervision, R.S.N.; project administration, R.A.V.; funding acquisition, L.A.L. All authors have read and agreed to the published version of the manuscript.

Funding: This research received no external funding.

Institutional Review Board Statement: The study was conducted in accordance with the Declaration of Helsinki and approved by the Ethics Committee, and the experiment was analyzed using the archival data of wrestlers (aged 27 to 47 years today).

Informed Consent Statement: Informed consent was obtained from all subjects involved in the study.

Data Availability Statement: Not Applicable.

Conflicts of Interest: The authors declare no conflict of interest.

References

1. Baca, A.; Dabnichki, P.; Hu, C.W.; Kornfeind, P.; Exel, J. Ubiquitous Computing in Sports and Physical Activity-Recent Trends and Developments. *Sensors* **2022**, *22*, 8370. [CrossRef] [PubMed]
2. Mikhaylova, I.V.; Petrova, M.A.; Bakulina, E.D. Digital transformation of chess training. *Teor. I Prakt. Fiz. Kult.* **2021**, *1*, 102–104.
3. Li, L.; Zhang, L.; Zhang, S. Using Artificial Intelligence for the Construction of University Physical Training and Teaching Systems. *J. Healthc. Eng.* **2021**, *29*, 3479208. [CrossRef] [PubMed]
4. Nagovitsyn, R.S.; Gibadullin, I.G.; Batsina, O.N.; Mokrushina, I.A. Forecasting the competitive performance of young athletes based on artificial intelligence technology. *Teor. I Prakt. Fiz. Kult.* **2023**, *2*, 27–30.
5. Serrano, M.; Khorsand, B.; Soldatos, J.; Troiano, E.; Neises, J.; Kranas, P.; Perakis, K.; Mamelli, A.; Elicegui, I.; Kyriazis, D.; et al. *Concepts and Design Thinking Innovation Addressing the Global Financial Needs: The INFINI-TECH Way Foundations*; Now Publishers: Boston, MA, USA; Delft, The Netherlands, 2023. [CrossRef]

6. Sahoo, S.; Lo, C.Y. Smart manufacturing powered by recent technological advancements: A review. *J. Manuf. Syst.* **2022**, *64*, 236–250. [CrossRef]
7. Biran, O.; Feder, O.; Moatti, Y.; Kiourtis, A.; Kyriazis, D.; Manias, G.; Mavrogiorgou, A.; Sgouros, N.M.; Barata, M.T.; Oldani, I.; et al. Policy CLOUD: A prototype of a cloud serverless ecosystem for policy analytics. *Data Policy* **2022**, *4*, E44. [CrossRef]
8. Wang, T.; Park, J. Design and implementation of intelligent sports training system for college students' mental health education. *Front. Psychol.* **2021**, *12*, 634978. [CrossRef] [PubMed]
9. Zhang, B.; Jin, H.; Duan, X. Physical education movement and comprehensive health quality intervention under the background of artificial intelligence. *Front. Public Health* **2022**, *10*, 947731. [CrossRef]
10. Kyriazis, D.; Autexier, S.; Boniface, M.; Engen, V.; Jimenez-Peris, R.; Jordan, B.; Jurak, G.; Kiourtis, A.; Kosmidis, T.; Lustrek, M.; et al. The CrowdHEALTH project and the Hollistic Health Records: Collective Wisdom Driving Public Health Policies. *Acta Inform Med.* **2019**, *27*, 369–373. [CrossRef]
11. Liu, H.; Zhu, X. Design of the Physical Fitness Evaluation Information Management System of Sports Athletes Based on Artificial Intelligence. *Comput. Intell. Neurosci.* **2022**, *30*, 1925757. [CrossRef]
12. Morhat, P.M. Artificial intelligence in the field of sports: Opportunities, directions and methods of involvement. *Teor. I Prakt. Fiz. Kult.* **2018**, *10*, 95–97.
13. Bobkova, E.N.; Parfianovich, E.V. Neural Networks for Forecasting and Modeling Training in Track-and Field Athletics. *Human. Sport. Med.* **2018**, *18*, 115–119. [CrossRef]
14. Dindorf, C.; Bartaguiz, E.; Gassmann, F.; Fröhlich, M. Conceptual Structure and Current Trends in Artificial Intelligence, Machine Learning, and Deep Learning Research in Sports: A Bibliometric Review. *Int. J. Environ. Res. Public Health* **2022**, *20*, 173. [CrossRef] [PubMed]
15. Koliada, M.G.; Belykh, S.I.; Bugayova, T.I.; Oleinik, O.S. Artificial intelligence method to detect psychological and pedagogical anomalies in physical education and sports activities. *Teor. I Prakt. Fiz. Kult.* **2021**, *9*, 66–69.
16. Barron, D.; Ball, G.; Robins, M.; Sunderland, C. Identifying playing talent in professional football using artificial neural networks. *J. Sports Sci.* **2020**, *38*, 1211–1220. [CrossRef] [PubMed]
17. Chikov, A.E.; Pavlov, E.A.; Egorov, N.A.; Medvedev, D.S.; Chikova, S.N.; Drobintsev, P.D. Artificial intelligence modeling of physiological parameters at anaerobic threshold. *Human. Sport. Med.* **2022**, *22* (Suppl. 2), 46–53.
18. Kakavas, G.; Malliaropoulos, N.; Pruna, R.; Maffulli, N. Artificial intelligence: A tool for sports trauma prediction. *Injury* **2020**, *51*, S63–S65. [CrossRef] [PubMed]
19. Zhang, J.; Qu, Q.; An, M.; Li, M.; Li, K.; Kim, S. Influence of Sports Biomechanics on Martial Arts Sports and Comprehensive Neuromuscular Control under the Background of Artificial Intelligence. *Contrast Media Mol. Imaging* **2022**, *10*, 9228838. [CrossRef]
20. Dhar, V. What Is the Role of Artificial Intelligence in Sports? *Big Data* **2017**, *5*, 173–174. [CrossRef]
21. Rodrigues, A.C.N.; Pereira, A.S.; Mendes, R.M.S.; Araújo, A.G.; Couceiro, M.S.; Figueiredo, A.J. Using Artificial Intelligence for Pattern Recognition in a Sports Context. *Sensors* **2020**, *20*, 3040. [CrossRef]
22. Minbaleev, A.V.; Titova, E.V. The Use of Artificial Intelligence in Sports and Its Legal Limitations. *Human. Sport. Med.* **2020**, *20* (Suppl. 2), 114–119.
23. Lu, G. Prediction Model and Data Simulation of Sports Performance Based on the Artificial Intelligence Algorithm. *Comput. Intell. Neurosci.* **2022**, *7*, 7238789. [CrossRef]
24. Yue, Q. Dynamic database design of sports quality based on genetic data algorithm and artificial intelligence. *Comput. Intell. Neurosci.* **2022**, *16*, 7473109. [CrossRef]
25. Xu, J. Prediction and Planning of Sports Competition Based on Deep Neural Network. *Comput. Intell. Neurosci.* **2022**, *8*, 1906580. [CrossRef]
26. Liu, H.; Gamboa, H.; Schultz, T. Sensor-Based Human Activity and Behavior Research: Where Advanced Sensing and Recognition Technologies Meet. *Sensors* **2023**, *23*, 125. [CrossRef] [PubMed]

sports

Article

A Muscle Load Feedback Application for Strength Training: A Proof-of-Concept Study

Lisa Noteboom [1,*], Anouk Nijs [1], Peter J. Beek [1], Frans C. T. van der Helm [2] and Marco J. M. Hoozemans [1]

[1] Department of Human Movement Sciences, Vrije Universiteit Amsterdam, Amsterdam Movement Sciences, 1081 BT Amsterdam, The Netherlands; a.nijs@vu.nl (A.N.); p.j.beek@vu.nl (P.J.B.); m.j.m.hoozemans@vu.nl (M.J.M.H.)

[2] Department of Biomechanical Engineering, Delft University of Technology, 2628 CD Delft, The Netherlands; f.c.t.vanderhelm@tudelft.nl

* Correspondence: l.noteboom@vu.nl; Tel.: +31-6-34-14-39-41

Abstract: Muscle overload injuries in strength training might be prevented by providing personalized feedback about muscle load during a workout. In the present study, a new muscle load feedback application, which monitors and visualizes the loading of specific muscle groups, was developed in collaboration with the fitness company Gymstory. The aim of the present study was to examine the effectiveness of this feedback application in managing muscle load balance, muscle load level, and muscle soreness, and to evaluate how its actual use was experienced. Thirty participants were randomly distributed into 'control', 'partial feedback', and 'complete feedback' groups and monitored for eight workouts using the automatic exercise tracking system of Gymstory. The control group received no feedback, while the partial feedback group received a visualization of their estimated cumulative muscle load after each exercise, and the participants in the complete feedback group received this visualization together with suggestions for the next exercise to target muscle groups that had not been loaded yet. Generalized estimation equations (GEEs) were used to compare muscle load balance and soreness, and a one-way ANOVA was used to compare user experience scores between groups. The complete feedback group showed a significantly better muscle load balance ($\beta = -18.9$; 95% CI [−29.3, −8.6]), adhered better to the load suggestion provided by the application (significant interactions), and had higher user experience scores for Attractiveness ($p = 0.036$), Stimulation ($p = 0.031$), and Novelty ($p = 0.019$) than the control group. No significant group differences were found for muscle soreness. Based on these results, it was concluded that personal feedback about muscle load in the form of a muscle body map in combination with exercise suggestions can effectively guide strength training practitioners towards certain load levels and more balanced cumulative muscle loads. This application has potential to be applied in strength training practice as a training tool and may help in preventing muscle overload.

Keywords: strength training; resistance training; feedback; injury; muscles; overload; overtraining

Citation: Noteboom, L.; Nijs, A.; Beek, P.J.; van der Helm, F.C.T.; Hoozemans, M.J.M. A Muscle Load Feedback Application for Strength Training: A Proof-of-Concept Study. *Sports* **2023**, *11*, 170. https://doi.org/10.3390/sports11090170

Academic Editors: Flávio Antônio de Souza Castro, Rui Miguel Simões de Azevedo and Rodrigo Zacca

Received: 6 July 2023
Revised: 7 August 2023
Accepted: 10 August 2023
Published: 5 September 2023

1. Introduction

Participation in fitness and strength training activities is becoming increasingly popular and brings along various health benefits [1,2]. Nevertheless, sedentary lifestyles remain a serious health issue worldwide [3]. In times of growing sedentary behavior and related adverse health problems, fitness and strength training activities can play a major role in improving public health worldwide, because they are readily accessible [3]. However, on the downside, there is a health risk related to fitness and strength training activities in the form of musculoskeletal injuries. In 2019, fitness injuries accounted for 17% (930,000) of all sports injuries in the Netherlands, ranking this sports domain second behind soccer (22%) [4]. The challenge is to stimulate people to engage in fitness and strength training and reap all the associated health benefits while reducing the risk of fitness-related musculoskeletal injuries.

In line with the widely used injury prevention model of van Mechelen et al. [5], the first steps towards prevention include gaining knowledge regarding fitness-related injuries and their etiology. Few studies have investigated fitness-related injuries in the general population [4,6]. However, more studies have been conducted on fitness-related injuries in more competitive sub-groups of fitness, such as CrossFit, powerlifting, and weightlifting [7–11]. Although these sub-disciplines differ from general fitness and strength training because they focus on specific exercises only, similar loading principles and objectives apply. Therefore, these sub-disciplines can provide meaningful information regarding fitness-related injuries. Based on the combined results reported in those studies, 58–59% of the injuries involved the muscles and tendons [6,7]. Most injuries were mild to moderate in severity, resulting in relatively short (training) time losses [6,8,9]. Regarding the injury onset (acute or overuse), overuse injury percentages varied between 25 and 54% of all injuries [4,6,8,9,11], which can be considered high compared to other sports [4]. A systematic review [12] reported 'overuse' and the 'frequent use of high loads' as two of the five main causes of injury and musculoskeletal pain in fitness. Moreover, two studies reported that the main injury cause mentioned by fitness practitioners was overload [6,13]. Overload can result from excessive training volumes or insufficient recovery time. In one study, the specific types of overload causing injury that were mentioned by practitioners included 'using too much weight', 'executing too many repetitions', or 'performing too many activities' [6]. Based on these findings, it seems that a large part of the fitness injury problem concerns muscle injuries that are often caused by overload or overuse or at least are perceived as resulting from those causes.

According to step three of van Mechelen et al.'s injury prevention model [5], a preventive measure is needed to prevent muscle overload. Currently, only 18% of all fitness participants train under the continuous supervision of a personal trainer or instructor [6]. Without expert supervision, strength training practitioners may not know the appropriate weights and number of repetitions to choose and may not be fully aware of which muscles they are training, which could lead to overloading of specific muscles. A potential solution might be to provide practitioners with feedback regarding their muscle load during their workout via a mobile application. However, to our knowledge, no such application has been developed, perhaps due to the technical difficulties involved in assessing muscle load, as well as various other issues that should be addressed. The first factor to consider is the modality of the feedback (e.g., visual, acoustic, or haptic). Visual feedback has been proven to be an effective means of performance enhancement in strength training [14] and is easy to provide via a mobile application. A second factor is the timing of the feedback, which can be concurrent, i.e., during the performance, or terminal, i.e., after the performance. Concurrent feedback may help with understanding a complex task, especially during the early phases of learning, whereas terminal feedback may lead to better learning in easy tasks or during later phases of learning [15]. A third factor is the content of the feedback, which must be relevant to the task and understandable to the user [15]. For the purpose of providing practitioners with feedback regarding their muscle load during their workout, we opted for visual, terminal feedback on muscle load after each exercise, personalized to their own capacity. This type of feedback could potentially help practitioners to avoid overloading their muscles, and instead load all muscles with an appropriate intensity that matches their capacity and results in achieving the desired training goal (e.g., hypertrophy). Additionally, apart from the perspective of injury prevention, practitioners might be more motivated to keep on exercising once they receive stimulating information during their workout in relation to their training goals [14].

In summary, muscle overload injuries form a problem in strength training and may be prevented by providing practitioners with feedback regarding their muscle load. As an initial step towards this goal, we developed and evaluated an innovative muscle load feedback application for strength training, with the overarching long-term aim of reducing injury risk by avoiding overload and overuse. In this study, we sought to determine if the feedback application in question can aid users to achieve a more balanced muscle load, by avoiding overloading certain muscles while underloading other muscles. A more balanced load may contribute to the prevention of overload injuries, which is an assumption that must be confirmed in future controlled trials with prospective longitudinal research designs. However, self-reported muscle soreness, which may be considered a precursor of muscle injuries [16], can be assessed in the short term. The primary aim of this proof-of-concept study was therefore to investigate if muscle load feedback can improve the muscle load balance, muscle load level, and muscle soreness of strength training practitioners, while the secondary aim was to evaluate the user experience and motivational effect of the application. In particular, it was hypothesized that muscle load feedback improves the muscle load balance and muscle soreness and has a positive effect on the user experience of strength training participants. If these hypotheses are confirmed, the feedback application may be further refined and validated.

2. Materials and Methods

2.1. Participants

The minimum sample size required to obtain an effect size of 0.25 with a power of 80% and a statistical significance level of 5% was calculated by G*Power 3 [17], assuming a mixed ANOVA design, since the literature regarding power analyses for Generalized Estimation Equations (GEEs), the statistical method adopted in the present study, was lacking. A minimum required sample size of 21 participants was found. To be certain of adequate power, 30 healthy participants (sex: 16 men, 14 women, age: 37 ± 14 years, strength training experience: 2.4 ± 2.4 years; mean ± standard deviation (SD)) were included in the present study. The inclusion criteria were a minimum age of 18 years and no musculoskeletal injuries at the start of the study. Recruitment was aimed at obtaining an as heterogenous sample as possible in terms of sex, age, and strength training experience, to achieve generalizable results. The participants were recruited through voluntary response sampling. The study was approved by the local ethics committee of the Faculty of Behavioural and Movement Sciences of the Vrije Universiteit Amsterdam (VCWE-2021-188). All participants provided written informed consent.

2.2. Muscle Load Feedback Application

The feedback application was developed in collaboration with the fitness company Gymstory [18]. Gymstory provides an automated gym exercise tracking solution, based on a sensor and mobile app (Gymstory, Amsterdam, The Netherlands). With this solution, the performed exercises, lifted masses, and number of performed repetitions and sets can be tracked and registered for each individual strength training athlete. Since the purpose of the study was to provide muscle load feedback, we extended the Gymstory app with additional information and functionalities to be able to estimate personalized muscle loads. To this end, multiple components had to be implemented. First of all, although Gymstory tracks the training volume and intensity by registering the repetitions and masses, the maximum masses that participants can lift for specific exercises must be known to personalize this intensity. Therefore, the option to add the one-repetition max (1RM) per exercise was implemented. Second, estimating the muscle load for an exercise requires knowledge of which muscle groups are contributing to that movement and to what extent. These muscle contributions can be estimated based on functional anatomical knowledge. A database that described primary and secondary muscle contributions for multiple exercises (Supplementary File S1) was implemented in the Gymstory app. Thirdly, an equation to estimate an athlete's personalized muscle load, based on recorded repetitions and masses, measured 1RMs, and estimated muscle contributions was added (Supplementary File S2). The equation is based on the assumption that when an exercise is performed for 3 sets of 10 repetitions at 70% of the 1RM, which corresponds to guidelines from the American College of Sports Medicine (ACSM) [19], the target load is achieved for a primary muscle, and half of the target load is achieved for a secondary muscle. More details and assumptions for the equation are presented in Supplementary File S2. This equation was implemented in the Gymstory app. It must be noted that the muscle load calculations remain to be validated. For the present study, muscle load estimates obtained by using this equation were deemed sufficient to examine if the concept of the feedback application works. If the concept proves successful, the accuracy of the calculations could be improved.

By employing the implemented components, a personalized muscle load could be estimated with the Gymstory app after each exercise. To provide clear and understandable feedback to the practitioners, the muscle load was visualized in the app as a colored muscle load body map (Figure 1A). This visualization represents the cumulative load that is updated after each exercise. At the start of a workout session, all muscle groups in the body map are white, and after each exercise, the colors of the corresponding primary and secondary muscle groups are updated, with light green indicating light to medium cumulative muscle load, dark green indicating the target muscle load, and yellow/orange/red indicating a potential risk of muscle overload. In addition to the muscle load body map, another functionality was added to the app, which provided practitioners with advice for subsequent exercises (including appropriate sets, repetitions, and weights) that targeted muscle groups that had not or had barely been loaded yet (Figure 1B). This feature rested on the assumption that a more balanced cumulative muscle load would prevent overloading of specific muscles and underloading of other muscles, considering a total body workout.

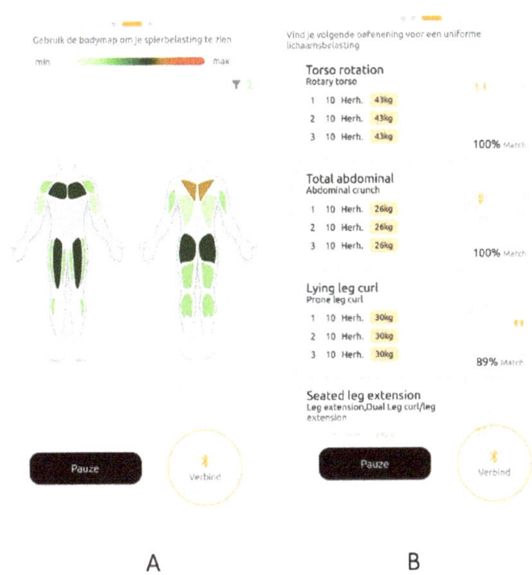

Figure 1. Screenshots of the Gymstory app with (**A**) the muscle load body map showing the cumulative muscle load per muscle by color and (**B**) a list with suggestions for subsequent exercises that target muscle groups that have not or had barely been loaded yet.

2.3. Procedures

The study design was a matched experimental study with three groups. A flow diagram of the study design is shown in Figure 2. The participants were first matched by an optimization method that minimized the variations in sex, age, and strength training experience, and subsequently randomly allocated them to one of three groups: control group, 'partial feedback' group, and 'complete feedback' group. All participants started with two intake sessions with an experienced fitness instructor in which their one-repetition maximum (1RM) was estimated by a submaximal test protocol for 18 selected exercises [20]. Submaximal 1RM testing involves a safer protocol than maximal 1RM testing in which the participant performs a set of repetitions with a submaximal weight until failure. The number of correctly executed repetitions and the lifted weight are used to estimate the 1RM based on a regression equation [20]. Next, all participants performed eight strength training sessions for four consecutive weeks (two per week). In each session, the participants were instructed to perform 8 exercises of their choice from the 18 preselected exercises. All participants were instructed to attempt to perform a full-body workout, loading all muscles evenly. Performed exercises, sets, repetitions, and weights were automatically tracked by the Gymstory sensor. Based on this data in combination with the measured 1RMs and functional anatomical knowledge of primary and secondary muscle contributions per exercise, the cumulative load on each of the muscle groups for each participant was individually estimated using the developed equation (Supplementary File S2). The three groups received the following feedback regarding their muscle load during the eight workouts:

- Control group: no feedback.
- Partial feedback: this group received a visualization of the estimated cumulative muscle load after each exercise via the Gymstory app in the form of the muscle load body map (Figure 1A).
- Complete feedback: this group also received the muscle load body map after each exercise, and additionally received a list with suggestions for subsequent exercises, targeting muscle groups that had not or had barely been loaded yet (Figure 1A,B).

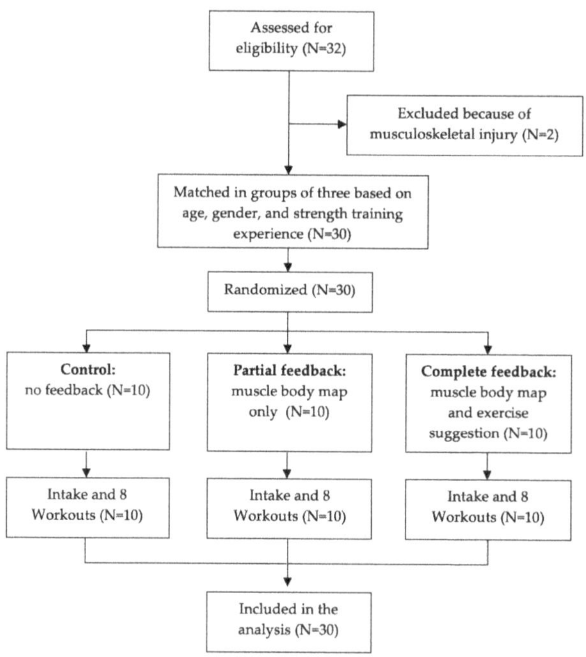

Figure 2. Flow diagram of the study design.

The morning after each workout, all participants received an online questionnaire with questions regarding their muscle soreness (Supplementary File S3) and were invited to score the soreness per muscle group by self-assessment on a visual analogue scale (VAS) (0–100 mm), with 0 indicating 'no soreness' and 100 indicating 'maximum soreness'. The participants were asked not to perform any other sports activities in the 24 h before or on the same day after the strength workout, in order to avoid muscle soreness from other sports at the time of the questionnaire. Furthermore, after the last workout, the user experience questionnaire (UEQ) was added to the questionnaires [21,22] to examine how the practitioners experienced the application. The UEQ consists of 26 pairs of opposing terms, each belonging to one of six scales: Attractiveness, Perspicuity, Efficiency, Dependability, Stimulation, and Novelty. The answers were rated on a 7-point Likert scale ranging from −3 (completely agree with the term on the left) to +3 (completely agree with the term on the right).

2.4. Statistical Analyses

A frequency analysis was performed to describe the characteristics of each group. After a workout, the total cumulative muscle load of a participant (calculated by the equation in Supplementary File S2) was presented as an array, with a percentage between 0 and 100+ for each muscle group. From this array, the SD was calculated as an indicator of cumulative muscle load balance. A lower SD meant a more even load distribution over all muscles, while a higher SD meant that, compared to the mean loading of the muscles, certain muscle groups were loaded extensively while other muscle groups were not. In addition, the mean of the cumulative muscle load was calculated after each session. For the muscle soreness, the mean was calculated for each session. The SD of the cumulative muscle load was expressed as a percentage of the mean cumulative muscle load of that session for normalization (i.e., the coefficient of variation; CV). The statistical analyses were performed in IBM SPSS Statistics 25. The cumulative muscle load balance, mean cumulative muscle load, and mean muscle soreness were compared between the three groups and eight workout numbers using Generalized Estimation Equations (GEEs) with an exchangeable working correlation structure [23]. GEEs were chosen as the regression technique because this technique can deal with correlated data (repeated measurements) and missing data (missing workouts). In the initial GEE regression model, the categorical variables group (type of feedback), workout number, and the interaction between group and workout number were included as predictor variables. If the parameter estimates showed that none or only a very few of the (in total 14) group × workout number interaction combinations were significant, and no clear interaction effect could be observed from the corresponding figure, the interaction term was removed from the model. The parameter estimates and their 95% confidence intervals (CIs) for the final GEE models were reported for all outcome parameters, indicating that an a priori α level of 0.05 was used to determine statistical significance. In all regression analyses, the control group and the first workout were always taken as a reference for computing the contrasts.

The UEQ scores for the six scales were compared between groups using a one-way ANOVA. Effect sizes r are reported, where $r = 0.1$ represents a small effect, $r = 0.3$ a medium effect, and $r = 0.5$ a large effect. In the case of a significant effect, post hoc analyses with Bonferroni correction were conducted. In addition, the scores were compared to the established benchmark values for the UEQ, which were constructed based on a dataset containing 9905 responses from 246 product evaluations [22]. This informs us not only about the differences in user experience between the different groups in our study, but also about how the user experience holds up compared to a large set of other products.

3. Results

In total, 30 participants completed 191 workouts, implying that 49 sessions were missed, e.g., due to illness. All participants completed at least four workouts. The participant characteristics per group are summarized in Table 1. An ANOVA on these data confirmed that the matching and randomization procedure was successful and that there were no between-group differences in sex ($p = 0.886$), age ($p = 0.934$), or strength training experience ($p = 0.963$) (Table 1).

Table 1. Participant characteristics per group and the p-values of the one-way ANOVAs comparing those characteristics between the groups.

Item	Control (n = 10)		Partial Feedback (n = 10)		Complete Feedback (n = 10)		p-Value
	n (%)	Mean ± SD	n (%)	Mean ± SD	n (%)	Mean ± SD	
Sex Male	5 (50%)		6 (60%)		5 (50%)		0.886
Female	5 (50%)		4 (40%)		5 (50%)		
Age (years)		38 ± 14		39 ± 15		36 ± 15	0.934
Strength training experience (years)		2.5 ± 2.6		2.2 ± 2.4		2.4 ± 2.5	0.963

3.1. Muscle Load

The group means of the normalized standard deviation of the cumulative muscle load, which was used as an indicator of muscle load balance, are presented for all workouts in Figure 3A. It can be observed from Figure 3 that, in general, the cumulative muscle load CV was lowest for the complete feedback group, followed by the partial feedback group and the control group, while the between-group differences remained reasonably constant over all workouts. The initial GEE model consisted of group, workout number, and their interaction as predictor variables of the outcome cumulative muscle load CV. This analysis revealed only one significant interaction term (from a total of 14 interaction terms between the categories of group and workout number), confirming that the differences between groups remained similar throughout the workouts. The interaction between group and workout number was therefore removed from the model and the final GEE model for cumulative muscle load CV (Table 2) showed that the cumulative muscle load CV was, for all workouts, significantly lower in the complete feedback group compared with the control group (β = −18.9; 95% CI [−29.3, −8.6]). The cumulative muscle load CV for the partial feedback group, although lower, was not significantly different from the control group (β = −6.8; 95% CI [−19.5, 6.0]). There were no significant differences in cumulative muscle load CVs between each of the workouts 2–8 and workout 1 (the reference workout), indicating that the cumulative muscle load CV did not change over time for each of the groups. These results indicated that a more balanced (estimated) muscle load was achieved with the complete feedback, but not with the partial feedback, compared with the control group.

In addition, it was examined whether the group, workout, or the interaction between group and workout influenced the mean cumulative muscle load. The corresponding regression analysis revealed multiple significant interactions (Table 2). In line with what can be observed in Figure 3B, these analyses showed that the mean cumulative muscle load increased significantly more from workout 1 to workout 6 in the control group than in the partial feedback group (β = −24.0, 95% CI [−47.9, −0.1]) and the complete feedback group (β = −24.3, 95% CI [−46.3, −2.2]). Also, between workouts 1 and 8, the cumulative muscle load increased significantly more in the control group than in the partial feedback group (β = −37.6, 95% CI [−62.4, −12.9]) and the complete feedback group (β = −26.3, 95% CI [−49.5, −3.0]). Figure 3B shows that while the mean cumulative muscle loads were similar in all groups at the initial workouts, the mean cumulative muscle load in the control group increased throughout the workouts, whereas the load in the feedback groups remained similar, explaining the observed significant interactions between group and workout numbers.

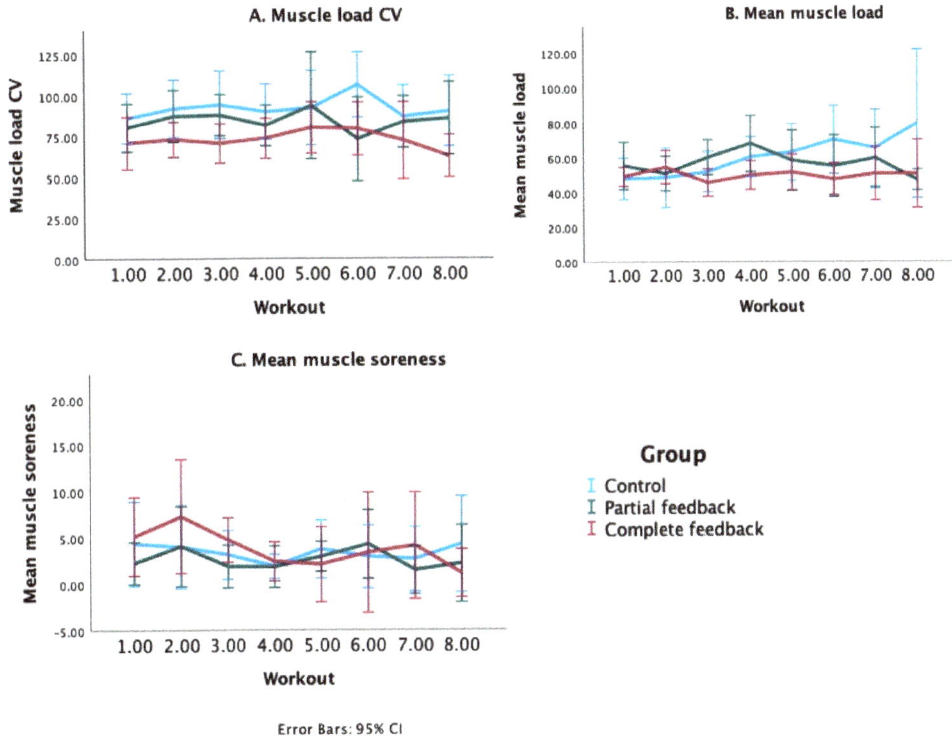

Figure 3. Group means of (**A**) the cumulative muscle load CV (coefficient of variation; indicator of muscle load balance), (**B**) the mean cumulative muscle load, (**C**) the mean muscle soreness, per workout. Error bars (95% confidence intervals (CI)) indicate the variation between participants.

Table 2. Parameter estimates (Beta) and their 95% confidence intervals (CIs) of the final Generalized Estimation Equation models for cumulative muscle load CV (coefficient of variation), mean cumulative muscle load, and mean muscle soreness.

Independent Variables		Cumulative Muscle Load CV (β and 95% CI)	Mean Cumulative Muscle Load (β and 95% CI)	Mean Muscle Soreness (β and 95% CI)
Group	Control [a]	0	0	0
	Partial	−6.8 (−19.5, 6.0)	7.5 (−7.3, 22.3)	−0.5 (−2.8, 1.8)
	Complete	−18.9 (−29.3, −8.6) *	1.3 (−9.5, 12.0)	0.9 (−1.8, 3.6)
Workout number	1 [a]	0	0	0
	2	5.0 (−4.4, 14.3)	0.7(−14.4, 15.7)	1.2 (−1.4, 3.8)
	3	5.1 (−3.9, 14.1)	4.0 (−7.8, 15.8)	−0.7 (−2.5, 1.2)
	4	2.5 (−5.0, 10.0)	12.5 (−4.8, 29.8)	−1.9 (−3.5,−0.2) *
	5	8.9 (−0.6, 18.4)	16.4 (−1.1, 33.9)	−1.0 (−2.4, 0.5)
	6	6.7 (−3.7, 17.1)	22.1 (0.9, 43.4) *	−0.3 (−2.1, 1.5)
	7	0.6 (−10.2, 11.3)	16.1(−0.9, 33.0)	−1.2 (−3.5, 1.0)
	8	−0.3 (−8.7, 8.1)	27.6 (4.7, 50.5) *	−1.1 (−3.5, 1.4)

Table 2. *Cont.*

Independent Variables		Cumulative Muscle Load CV (β and 95% CI)	Mean Cumulative Muscle Load (β and 95% CI)	Mean Muscle Soreness (β and 95% CI)
Interaction group * workout number	Control*1 [a]		0	
	Control*2 [a]		0	
	Control*3 [a]		0	
	Control*4 [a]		0	
	Control*5 [a]		0	
	Control*6 [a]		0	
	Control*7 [a]		0	
	Control*8 [a]		0	
	Partial*1 [a]		0	
	Partial*2		−5.2 (−24.0, 13.7)	
	Partial*3		0.7 (−15.5, 17.0)	
	Partial*4		0.2 (−19.4, 19.8)	
	Partial*5		−15.2 (−37.3, 7.0)	
	Partial*6		−24.0 (−47.9, −0.1) *	
	Partial*7		−13.5 (−33.9, 6.8)	
	Partial*8		−37.6 (−62.4, −12.9) *	
	Complete*1 [a]		0	
	Complete*2		4.8 (−13.4, 22.9)	
	Complete*3		−7.6 (−21.3, 6.1)	
	Complete*4		−11.9(−30.1, 6.4)	
	Complete*5		−14.5 (−34.2, 5.3)	
	Complete*6		−24.3 (−46.3, −2.2) *	
	Complete*7		−14.3(−33.5, 4.9)	
	Complete*8		−26.3 (−49.5, −3.0) *	

* Refers to an interaction term. [a] Refers to a reference group.

3.2. Muscle Soreness

It was investigated if the mean muscle soreness was influenced by the group, workout number, or interaction between group and workout number (Figure 3C). Significant interactions between group and workout were not observed, and the final regression model without the interactions only showed a significantly lower mean soreness at workout 4 compared with workout 1 (β = −1.9, 95% CI [−3.5, −0.2]) for all groups (Table 2).

3.3. User Experience

The group mean scores for the six scales of the User Experience Questionnaire (UEQ) after the last workout for all three groups are presented in Figure 4. In the background of the figure, the benchmark scores for the different scales are presented. As can be observed in this figure, the scores ranged between bad and above average for the control group and the partial feedback group, whereas the scores were above average to excellent for all scales for the complete feedback group.

The one-way ANOVA revealed a significant effect of group on the scores of all six scales (Table 3). Post hoc analyses with Bonferroni correction showed that the complete feedback group rated the app significantly better than the control group on Attractiveness ($p = 0.036$), Stimulation ($p = 0.031$), and Novelty ($p = 0.019$), and significantly higher than the partial feedback group on Dependability ($p = 0.019$), although this difference was also nearly significant for Perspicuity ($p = 0.051$), Efficiency ($p = 0.051$), and Novelty ($p = 0.053$). There were no significant differences between the control group and the partial feedback group.

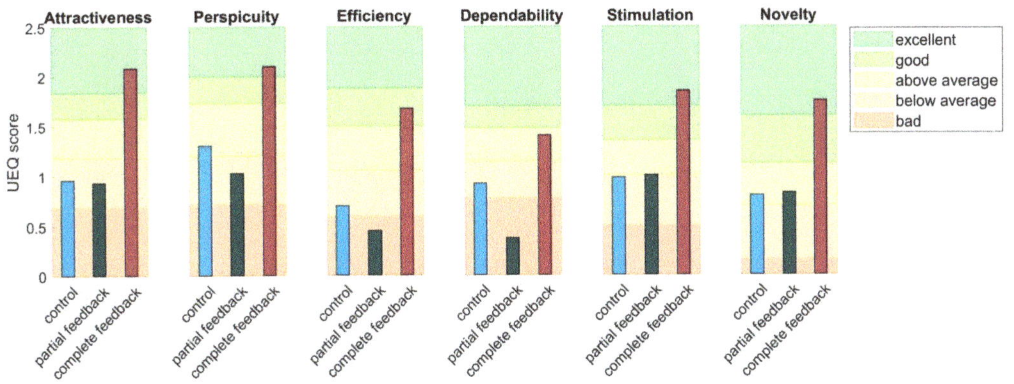

Figure 4. Overview of the mean scores on the different scales (subplots) for the different groups (colors) at the last training session. Shaded areas represent the benchmark values of <25%, 25–50%, 50–75%, 75–90%, and >90% performance, representing bad, below average, above average, good, and excellent performance, respectively.

Table 3. Results of the one-way ANOVA comparing the user experience questionnaire (UEQ) scores on each scale between groups, as well as the post hoc analyses.

Scale	Post Hoc Comparisons	Mean Difference	Standard Error	df	*F*-Value	*p*-Value	Effect Size *r*
Attractiveness				2, 26	4.465	0.022 *	0.506
	Control–Partial	−0.080	0.430			1.000	
	Control–Complete	−1.127	0.418			0.036 *	
	Partial–Complete	−1.047	0.430			0.066	
Perspicuity				2, 26	3.826	0.035 *	0.477
	Control–Partial	0.161	0.377			1.000	
	Control–Complete	−0.800	0.367			0.116	
	Partial–Complete	−0.961	0.377			0.051	
Efficiency				2, 26	3.810	0.035 *	0.476
	Control–Partial	0.200	0.462			1.000	
	Control–Complete	−0.975	0.449			0.118	
	Partial–Complete	−1.175	0.462			0.051	
Dependability				2, 26	4.403	0.023 *	0.503
	Control–Partial	0.500	0.332			0.430	
	Control–Complete	−0.483	0.323			0.439	
	Partial–Complete	−0.983	0.331			0.019 *	
Stimulation				2, 26	4.402	0.023 *	0.503
	Control–Partial	−0.136	0.325			1.000	
	Control–Complete	−0.875	0.316			0.031 *	
	Partial–Complete	−0.739	0.325			0.094	
Novelty				2, 26	5.170	0.013 *	0.533
	Control–Partial	−0.117	0.329			1.000	
	Control–Complete	−0.950	0.320			0.019 *	
	Partial–Complete	−0.833	0.329			0.053	

* $p < 0.05$.

4. Discussion

The primary aim of the present study was to investigate if the use of a newly conceived muscle load feedback application could effectively improve the cumulative muscle load balance, muscle load level, and muscle soreness balance in strength training practitioners during total body workouts. The results revealed that practitioners who received feedback in the form of a body map and exercise suggestions during their workout achieved a more balanced cumulative muscle load compared with practitioners who did not receive any feedback, as opposed to practitioners who only received the body map. Moreover, it was found that the mean cumulative muscle load was similar for all groups in the first workouts, but the mean load in the control group increased throughout the workouts while the load remained the same in the feedback groups. The mean muscle soreness was similar for all groups and all workouts, implying that no evidence was found that the feedback can decrease muscle soreness.

The secondary aim of the study was to evaluate the subjective experience with the application. User experience, as assessed using the UEQ and compared with the benchmark scores, showed that the complete feedback group rated the application good to excellent on all scales. Comparisons between groups showed that the rated scores were better for the complete feedback group than the other two groups on all scales.

The feedback provided in the form of a body map and exercise suggestion effectively improved the cumulative muscle load balance of the participants, indicating that participants adhered to the feedback provided by the app. To verify that the feedback was used, participants were questioned after each workout on what they had based their exercise choice on, and these results showed that nine out of ten participants in the complete feedback group had typically used the list of suggestions for subsequent exercises (regularly combined with the body map, and often combined with their own preference), and only one participant typically based the choice on his/her own preference or on the body map alone. In line with these results, high scores were found for all items of the user experience questionnaire in the complete feedback group, and 100% of the participants indicated that they would like to use the feedback during their future workouts, confirming that this type of feedback was perceived to be useful and valuable.

However, the feedback that consisted of the body map only, was not found to be significantly effective in improving participants' cumulative muscle load balance. A potential explanation for this finding is that even when participants know which muscle groups they need to train (because these muscle groups appear white or light green on the body map), they do not know exactly which exercises to choose to target these muscle groups. Moreover, the questionnaires showed that 60% of the participants in the partial feedback group typically used the body map to select their next exercise, but 40% made the choice based on their own preference or another strategy. The fact that a substantial portion (40%) of the participants in the partial feedback group did not use the feedback could also be a reason for the lack of effect on the muscle load balance. To further examine this theory, we compared the average cumulative muscle load CVs between participants in the partial feedback group who did and did not use the body map. Interestingly, the mean cumulative muscle load CV was much smaller (76.8% of the mean cumulative muscle load) for the participants who did use the body map to select their next exercise, compared with participants who did not (97.1% of the mean cumulative muscle load). This finding highlights that feedback consisting of the body map alone could potentially aid in improving the cumulative muscle load balance, but not for all practitioners, either because some practitioners may prefer other methods to select their exercises or perhaps because they do not know how to properly use this type of feedback. In line with this, the user experience scores were lower in the partial feedback group than in the complete feedback group. This shows that the content of this feedback can be improved upon in terms of relevance and/or understandability [15].

The mean cumulative muscle loads increased more throughout the workouts for the control group compared with the feedback groups. This probably occurred because the feedback groups received feedback with a load level that was determined during the intake sessions and not adjusted during the relatively short (four-week) period of the study, and participants in the feedback groups seemed to adhere to these load levels. This result suggests that this type of feedback can be effective in guiding strength training practitioners towards a certain load level. Therewith, the application could potentially aid in preventing practitioners from training too heavily (and thereby presumably prevent overload injuries). The participants that did not receive feedback increased their loads, which could constitute a risk for overtraining according to the ACSM [19]. However, it must be noted that, especially when this feedback application is applied for longer training periods, practitioners are expected to increase in strength, implying that the load prescriptions should also increase over time (meaning we cannot exclude that the control group increased their load appropriately). It is very important that appropriate load increases are determined carefully. For instance, regular (submaximal) 1RM tests could be conducted to update the 1RM values in the application. Another option that may be more applicable, is that practitioners regularly evaluate if they can perform one or two repetitions more than the prescribed number, in which case the load prescriptions in the app could be increased with 2–10% according to ACSM guidelines [19].

Although similar studies regarding cumulative muscle load feedback are lacking in the literature, the present findings can be compared to those of studies that applied different forms of feedback during strength training, including feedback on barbell velocities and body kinematics and kinetics. In line with our findings, Weakley et al. [24] and Wilson et al. [14] found that visual barbell-velocity feedback improved motivation and competitiveness of strength training participants. Furthermore, Keller et al. [25] found that kinematic jump height feedback led to immediate improvement in jump performance. This agrees with the immediate improvement in muscle load balance for the complete feedback group observed in the first workout of the present study. However, when the feedback was removed in the study of Keller et al. [25], all improvements in jump performance were lost. Similarly, another feedback study found that the strength training performance returned to pre-feedback level when the feedback was removed [26]. Although this was not specifically investigated in the present study, the immediate improvement observed in the complete feedback group with respect to the control group at the first workout and absence of clear further improvements in the subsequent workouts may suggest there is no learning effect and people will depend on the application to obtain the positive effects. Strength training participants would therefore be advised to continue using the app if made available in practice.

Whereas typical methods to estimate muscle loads in human movement science studies include electromyography (EMG) or motion capture combined with musculoskeletal modeling [27], a new muscle load estimation method was chosen for the present study. A clear advantage of this new method compared to the more typical methods is that it can easily be applied in the gym, as no sensors or markers need to be attached to the athlete's body. However, the validity of this new method has not been investigated yet, and it is expected that the muscle load estimations deviate from the actual muscle loads for multiple reasons. Firstly, the muscle load was estimated based on rough estimates of the muscle contributions per exercise, subdivided into either primary (100% contribution) or secondary (50% contribution) muscles. This subdivision in only two options is most likely an oversimplification, since muscles could also contribute with other percentages. Secondly, individual differences in anatomy and movement execution were not included in the estimation. It is important for future use of the application that the accuracy of the estimations is tested and further improved. Regarding the first point, the muscle force estimation could be improved if more detailed muscle force contributions are determined, for instance, by estimating the muscle forces with musculoskeletal modeling with or without EMG for multiple participants and multiple exercises, and using the mean estimated contributions to replace the current functional anatomy-based estimations. Regarding the second point, real-time measurements would be required to include individual anthropometry and movement executions in the muscle load estimation. Camera-based motion capture combined with a fast-computing musculoskeletal model is a recent development that may be a potential option to integrate within the feedback application in the future [28].

In addition to improving the accuracy of the muscle load calculations, other adjustments could be made to improve the applicability of the feedback application. In the present study, muscle load advice was always based on a full body workout, a fixed load percentage of 70% of the 1RM, and a fixed number of sets and repetitions. It would be valuable if the application was expanded with multiple training options, so that the muscle load advice could be adjusted based on the final training goal of the athlete (e.g., hypertrophy and speed). For instance, athletes should be able to choose the desired 1RM level, and whether to train a specific body part or the full body. Moreover, future research could focus on further optimizing the load advice based on personal characteristics including gender, age, and experience.

The mean muscle soreness was not lower in the feedback groups than in the control group. However, the mean muscle soreness was found to be very low for all groups (with a score of 3 out of 100 on average). Moreover, muscle soreness cannot be translated to muscle overload injuries directly, and it is therefore important that the application is evaluated for a longer period wherein injuries are monitored prospectively as well to determine if the feedback can be effective for injury prevention. In addition, it must be noted that muscle soreness was self-assessed and based on subjective perceptions, and it may therefore be difficult to compare the amount of muscle soreness between individuals and between groups.

Multiple strengths and limitations apply to the present study. One of the strengths is that the studied feedback application can be directly applied in the sport practice and is currently already being used by Gymstory to provide muscle load feedback to their users. Another strength pertains to the innovative characteristics of the study, which, to our knowledge, is the first study to develop and evaluate a muscle load feedback application for strength training. On the other hand, the study suffered from some noteworthy limitations. Firstly, the sample size of the study was relatively small with 10 participants per group, especially because substantial differences between individuals were observed (as can be deducted from the substantial size of the error bars in Figure 3 and the difference in reported adherence within the partial feedback group). Secondly, recruitment was performed through voluntary response sampling, which may have led to self-selection bias. It is possible that the participants that volunteered were already more interested in receiving feedback during strength training than people that did not volunteer. Thirdly, we did not include a group who only received the exercise suggestion. Therefore, we cannot conclude whether the better muscle balance and user experience scores found for the complete feedback group were due to the combination of body map and exercise suggestion, or mainly due to the exercise suggestion. The absence of improvements in the partial feedback group may point at the second possibility.

5. Conclusions

- Feedback regarding the personal muscle load of strength training participants provided in the form of a muscle body map and exercise suggestion can effectively aid in achieving a more balanced cumulative muscle load, which may decrease the risk of overloading certain muscles while underloading other muscles.
- Feedback regarding the muscle load in the form of a body map and suggested exercises was perceived to be valuable and stimulating.
- Feedback regarding the muscle load can effectively guide strength training participants towards a certain load level, presumably helping to maximize training effects without getting injured.
- Feedback regarding the muscle load does not change muscle soreness.
- Future research can improve the accuracy of the muscle load estimation.
- A longitudinal study with a longer follow-up period is needed to investigate if the feedback application can effectively prevent muscle injuries.

6. Practical Application

The proof-of-concept feedback application can successfully influence the cumulative muscle load of strength training practitioners. This highlights the potential of this type of feedback as a training tool. Future research should improve and validate the application and investigate if it can successfully prevent muscle overload injuries.

Supplementary Materials: The following supporting information can be downloaded at: https://www.mdpi.com/article/10.3390/sports11090170/s1, File S1: Primary and secondary muscle contributions; File S2: Muscle load estimation; File S3: Muscle soreness questionnaire.

Author Contributions: Conceptualization, L.N., A.N., P.J.B., F.C.T.v.d.H. and M.J.M.H.; methodology, L.N., A.N., P.J.B., F.C.T.v.d.H. and M.J.M.H.; formal analysis, L.N. and A.N.; investigation, L.N. and A.N.; writing—original draft preparation, L.N. and A.N.; writing—review and editing, L.N., A.N., P.J.B., F.C.T.v.d.H. and M.J.M.H.; supervision, P.J.B., F.C.T.v.d.H. and M.J.M.H.; project administration, P.J.B., F.C.T.v.d.H. and M.J.M.H.; funding acquisition, P.J.B., F.C.T.v.d.H. and M.J.M.H. All authors have read and agreed to the published version of the manuscript.

Funding: This research was funded by the Dutch Research Council (NWO), Domain Applied and Engineering Sciences, grant number P16–28 Project 4.

Institutional Review Board Statement: The study was approved by the local ethics committee of the Faculty of Behavioural and Movement Sciences of the Vrije Universiteit Amsterdam (VCWE-2021-188).

Informed Consent Statement: Written informed consent was obtained from all subjects involved in the study.

Data Availability Statement: The data presented in this study are available on request from the corresponding author. The data are not publicly available due privacy reasons.

Acknowledgments: The authors would like to thank Mathieu Wernsen, Lars van der Wal, and Michael Fahim for their technical support, Marielle Spekreijse for conducting the 1RM measurements, and the VU sport center for the use of their gym for the measurements.

Conflicts of Interest: The company Gymstory, who integrated our feedback proposal into their mobile application for this study, may benefit from the results of this study. The authors declare that the company Gymstory was not involved in the design of the study; the collection, analysis, or interpretation of the data; or in the writing of the manuscript.

References

1. Lidmaatschappen en Sportdeelname NOC*NSF over 2019. NOC*NSF Afdeling Sportparticipatie. Available online: www.nocnsf.nl (accessed on 21 August 2023).
2. Maestroni, L.; Read, P.; Bishop, C.; Papadopoulos, K.; Suchomel, T.J.; Comfort, P.; Turner, A. The Benefits of Strength Training on Musculoskeletal System Health: Practical Applications for Interdisciplinary Care. *Sports Med.* **2020**, *50*, 1431–1450. [CrossRef] [PubMed]
3. Park, J.H.; Moon, J.H.; Kim, H.J.; Kong, M.H.; Oh, Y.H. Sedentary Lifestyle: Overview of Updated Evidence of Potential Health Risks. *Korean J. Fam. Med.* **2020**, *41*, 365–373. [CrossRef]
4. *Sportblessures in Nederland. Cijfers 2019*; VeiligheidNL: Amsterdam, The Netherlands, 2019.
5. Van Mechelen, W.; Hlobil, H.; Kemper, H.C.G. Incidence, Severity, Aetiology and Prevention of Sports Injuries. A review of concepts. *Sports Med.* **1992**, *14*, 82–99. [CrossRef]
6. Kemler, E.; Noteboom, L.; van Beijsterveldt, A.-M. Characteristics of Fitness-Related Injuries in The Netherlands: A Descriptive Epidemiological Study. *Sports* **2022**, *10*, 187. [CrossRef]
7. Szeles, P.R.D.Q.; da Costa, T.S.; da Cunha, R.A.; Hespanhol, L.; Pochini, A.D.C.; Ramos, L.A.; Cohen, M. CrossFit and the Epidemiology of Musculoskeletal Injuries: A Prospective 12-Week Cohort Study. *Orthop. J. Sports Med.* **2020**, *8*, 1147–1153. [CrossRef] [PubMed]
8. Moran, S.; Booker, H.; Staines, J.; Williams, S. Rates and risk factors of injury in CrossFit: A prospective cohort study. *J. Sports Med. Phys. Fit.* **2017**, *57*, 1147–1153. [CrossRef]
9. Keogh, J.; Hume, P.A.; Pearson, S. Retrospective Injury Epidemiology of One Hundred One Competitive Oceania Power Lifters: The Effects of Age, Body Mass, Competitive Standard, and Gender. *J. Strength Cond. Res.* **2006**, *20*, 672–681. [CrossRef]
10. Abouammoh, N.; Aldebeyan, W.; Aljadaan, B.; Dous, A.; Almashouq, M.; Alsubaie, A.; Alshahrani, A.; Alghamdi, H.; Alsoliman, M. Injury rate among CrossFit participants and its associated risk factors in Riyadh: A cross-sectional study. *Saudi J. Sports Med.* **2021**, *21*, 45–50. [CrossRef]
11. Reichel, T.; Mitnacht, M.; Fenwick, A.; Meffert, R.; Hoos, O.; Fehske, K. Incidence and characteristics of acute and overuse injuries in elite powerlifters. *Cogent Med.* **2019**, *6*, 1588192. [CrossRef]

12. Bonilla, D.A.; Cardozo, L.A.; Vélez-Gutiérrez, J.M.; Arévalo-Rodríguez, A.; Vargas-Molina, S.; Stout, J.R.; Kreider, R.B.; Petro, J.L. Exercise selection and common injuries in fitness centers: A systematic integrative review and practical recommendations. *Int. J. Environ. Res. Public Health* **2022**, *19*, 12710. [CrossRef]
13. *Sportblessures in Nederland. Cijfers 2017*; VeiligheidNL: Amsterdam, The Netherlands, 2017.
14. Wilson, K.M.; Helton, W.S.; de Joux, N.R.; Head, J.R.; Weakley, J.J.S. Real-time quantitative performance feedback during strength exercise improves motivation, competitiveness, mood, and performance. *Proc. Hum. Factors Ergon. Soc. Annu. Meet.* **2017**, *61*, 1546–1550. [CrossRef]
15. Sigrist, R.; Rauter, G.; Riener, R.; Wolf, P. Augmented visual, auditory, haptic, and multimodal feedback in motor learning: A review. *Psychon. Bull. Rev.* **2012**, *20*, 21–53. [CrossRef] [PubMed]
16. Baumert, P.; Lake, M.J.; Stewart, C.E.; Drust, B.; Erskine, R.M. Genetic variation and exercise-induced muscle damage: Implications for athletic performance, injury and ageing. *Eur. J. Appl. Physiol.* **2016**, *116*, 1595–1625. [CrossRef]
17. Faul, F.; Erdfelder, E.; Lang, A.G.; Buchner, A. G* Power 3: A flexible statistical power analysis program for the social, behavioral, and biomedical sciences. *Behav. Res. Methods* **2007**, *39*, 175–191. [CrossRef]
18. Gymstory. Available online: https://www.yourgymstory.com (accessed on 21 August 2023).
19. American College of Sports Medicine. American College of Sports Medicine position stand. Progression models in resistance training for healthy adults. *Med. Sci. Sports Exerc.* **2009**, *41*, 687–708. [CrossRef]
20. Mayhew, J.L.; Ball, T.E.; Arnold, M.D.; Bowen, J.C. Relative muscular endurance performance as a predictor of bench press strength in college men and women. *J. Appl. Sport Sci. Res.* **1992**, *6*, 200–206.
21. Laugwitz, B.; Held, T.; Schrepp, M. Construction and Evaluation of a User Experience Questionnaire. In *HCI and Usability for Education and Work*; Holzinger, A., Ed.; Springer: Berlin/Heidelberg, Germany, 2008; pp. 63–76. ISBN 978-3-540-89349-3. [CrossRef]
22. Schrepp, M.; Thomaschewski, J.; Hinderks, A. Construction of a benchmark for the user experience questionnaire (UEQ). *Int. J. Interact. Multimed. Artif. Intell.* **2017**, *4*, 40–44. [CrossRef]
23. Liang, K.Y.; Zeger, S.L. Regression analysis for correlated data. *Annu. Rev. Public Health* **1993**, *14*, 43–68. [CrossRef] [PubMed]
24. Weakley, J.J.; Wilson, K.M.; Till, K.; Read, D.B.; Darrall-Jones, J.; Roe, G.A.; Phibbs, P.J.; Jones, B. Visual feedback attenuates mean concentric barbell velocity loss and improves motivation, competitiveness, and perceived workload in male adolescent athletes. *J. Strength Cond. Res.* **2019**, *33*, 2420–2425. [CrossRef] [PubMed]
25. Keller, M.; Lauber, B.; Gehring, D.; Leukel, C.; Taube, W. Jump performance and augmented feedback: Immediate benefits and long-term training effects. *Hum. Mov. Sci.* **2014**, *36*, 177–189. [CrossRef]
26. Chalker, W.J.; Shield, A.J.; Opar, D.A.; Rathbone, E.N.; Keogh, J.W. Effect of acute augmented feedback on between limb asymmetries and eccentric knee flexor strength during the Nordic hamstring exercise. *PeerJ* **2018**, *6*, e4972. [CrossRef] [PubMed]
27. Chiu, L.Z. Biomechanical methods to quantify muscle effort during resistance exercise. *J. Strength Cond. Res.* **2018**, *32*, 502–513. [CrossRef] [PubMed]
28. Uhlrich, S.D.; Falisse, A.; Kidziński, Ł.; Muccini, J.; Ko, M.; Chaudhari, A.S.; Hicks, J.L.; Delp, S.L. OpenCap: 3D human movement dynamics from smartphone videos. *bioRxiv* **2022**. [CrossRef]

sports MDPI

Article

Velocity-Based Strength Training: The Validity and Personal Monitoring of Barbell Velocity with the Apple Watch

Basil Achermann [1,2], Katja Oberhofer [1], Stephen J. Ferguson [2] and Silvio R. Lorenzetti [1,2,*]

1 Section Performance Sport, Swiss Federal Institute of Sport Magglingen (SFISM), 2532 Magglingen, Switzerland; basil.achermann@baspo.admin.ch (B.A.)
2 Institute for Biomechanics, ETH Zurich, 8092 Zurich, Switzerland
* Correspondence: silvio.lorenzetti@baspo.admin.ch

Abstract: Velocity-based training (VBT) is a method to monitor resistance training based on measured kinematics. Often, measurement devices are too expensive for non-professional use. The purpose of this study was to determine the accuracy and precision of the Apple Watch 7 and the Enode Pro device for measuring mean, peak, and propulsive velocity during the free-weighted back squat (in comparison to Vicon as the criterion). Velocity parameters from Vicon optical motion capture and the Apple Watch were derived by processing the motion data in an automated Python workflow. For the mean velocity, the barbell-mounted Apple Watch (r = 0.971–0.979, SEE = 0.049), wrist-worn Apple Watch (r = 0.952–0.965, SEE = 0.064) and barbell-mounted Enode Pro (r = 0.959–0.971, SEE = 0.059) showed an equal level of validity. The barbell-mounted Apple Watch (Vpeak: r = 0.952–0.965, SEE = 0.092; Vprop: r = 0.973–0.981, SEE = 0.05) was found to be the most valid for assessing propulsive and peak lifting velocity. The present results on the validity of the Apple Watch are very promising, and may pave the way for the inclusion of VBT applications in mainstream consumer wearables.

Keywords: resistance training; velocity-based training; barbell kinematics; sensor technology; inertial measurement unit (IMU); wearables

check for updates

Citation: Achermann, B.; Oberhofer, K.; Ferguson, S.J.; Lorenzetti, S.R. Velocity-Based Strength Training: The Validity and Personal Monitoring of Barbell Velocity with the Apple Watch. *Sports* **2023**, *11*, 125. https://doi.org/10.3390/sports11070125

Academic Editors: Flávio Antônio de Souza Castro, Rui Miguel Simões de Azevedo and Rodrigo Zacca

Received: 24 May 2023
Revised: 14 June 2023
Accepted: 21 June 2023
Published: 23 June 2023

1. Introduction

Resistance training (RT), which involves working against an external load to increase muscle mass, maximal strength, and power output, has been shown to offer a range of benefits for both athletes and recreational exercisers [1–4]. Various methods for monitoring training intensity have been introduced, including those based on subjective questionnaires and user ratings, as well as objective measures based on the heaviest moveable weight, i.e., one repetition maximum (1RM), or velocity data, also named velocity-based training (VBT) [5,6]. While linear position transducers (LPT) are reliable [7] and considered the gold standard for measuring barbell velocity, they are often too expensive for non-professional use [8–10]. Recent studies have therefore investigated the validity and reliability of more affordable technologies, such as cameras, smartphone apps, and wearables designed specifically for VBT [11–13]. However, a specific area of interest that has not been sufficiently explored is the potential use of consumer wearables capable of measuring barbell velocity for VBT, while also providing additional features (i.e., sleep tracking). Despite the growing popularity of these devices among athletes and recreational exercisers, their effectiveness and accuracy in this specific context remain largely unverified [14].

Wearable devices equipped with inertial measurement unit (IMU) sensors have gained popularity in the sports sciences community due to their convenience and accessibility [15]. Harnessing the potential of a smartwatch for VBT measurements could provide a practical and affordable method for athletes and recreational exercisers to monitor their training intensity and performance. The integration of VBT with health and fitness applications could further enable the development of personalized training regimes, better tracking of fitness progress, and enhanced fitness outcomes.

However, IMUs are known to exhibit random errors [16]. Thus, the validity of commercially available IMUs for VBT monitoring is still discussed in the field. Multiple studies have assessed the validity of different IMU-based VBT devices such as the Beast Sensor (Beast Technologies S.r.l., Brescia, Italy), VmaxPro (BM Sports Technology GmbH, Magdeburg, Germany) and Push Band 2 (Whoop, Boston, MA, USA) [11,12,17–19] in a variety of free weight and Smith machine-supported barbell exercises. No definitive conclusion has yet been drawn, thus highlighting the need for further investigation. It remains to be validated whether consumer wearables such as smartwatches, equipped with IMU sensors and exemplified by the Apple Watch (Apple Inc., Cupertino, CA, USA) [20,21], have the capacity to accurately measure barbell kinematics during free-weighted back squats.

To address this research gap, the primary goal of the present study was to validate the Apple Watch as a tool for measuring barbell kinematics in VBT against data from optical motion capture as gold standard. We assessed its performance during a VBT protocol on healthy, recreational RT athletes by comparing the resulting velocity data of the free-weighted back squat from two different watch positions (i.e., wrist and barbell), compared with Vicon (Vicon 3D Motion Systems, Oxford, UK) as a criterion [22]. As a secondary objective and for a comprehensive and practically oriented assessment, we further compared the results from the Apple Watch with a commercially available IMU-based VBT device, namely the Enode Pro device (Blaumann and Meyer Sports Technology UG, Magdeburg, Germany), formerly known as Vmaxpro [12,19,23,24]. Our aim was to provide scientifically robust, praxis-oriented insights into the validity of the Apple Watch for monitoring barbell kinematics during VBT, given its widespread popularity as a consumer electronic device and the growing interest in VBT among athletes and recreational exercisers.

2. Materials and Methods

The study protocol complied with the Declaration of Helsinki for Human Experimentation and was approved by the regional ethics committee (Kantonale Ethikkommission Bern, Nr: 2021-00403). The study design was developed to assess the validity of the Apple Watch 7′s IMU sensor during free-weighted back squats in healthy volunteers. The methodology employed in this study was similar to previous validation studies [8,11]. All participants gave written signed consent prior to any data acquisition.

2.1. Participants

The participants in the present study were 22 recreationally active RT athletes with at least 3 years of experience in performing free-weighted back squats (n = 12M/10F, age = 29.1 ± 5.2, height = 1.65 ± 0.4 m; body mass = 77.5 ± 12.6 kg; back squat 1RM = 134 ± 32 kg). Written informed consent was obtained from each participant prior to data acquisition. All participants were healthy and regularly physically active.

2.2. Study Design

Subjects reported to the lab for one testing session, where they performed back squats between 45 to 100 percent of their one-repetition maximum (1RM) at maximum voluntary lifting velocity, according to VBT guidelines [25]. The testing protocol comprised a warm-up, the assessment of each individual's load–velocity profile, an actual 1RM test, and a set performed to exhaustion. Details of the testing protocol are shown in Figure 1 and further outlined in the following paragraph.

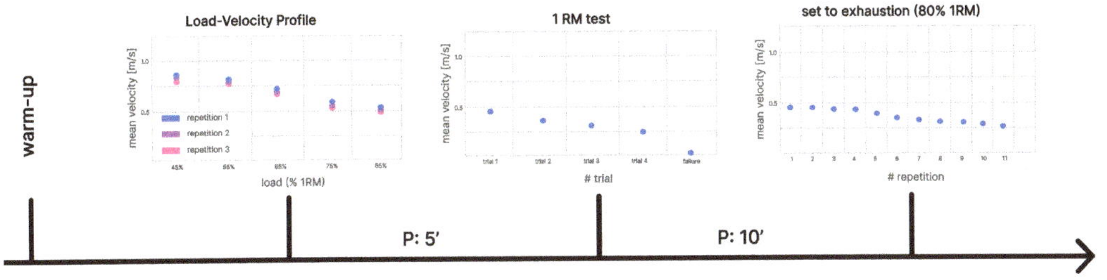

Figure 1. Visualization of testing protocol. It comprised a warm-up, the assessment of each individual's load–velocity profile, an actual 1RM test, and a set performed to exhaustion.

Each participant performed a 10 min individual warm-up according to their standard training routine, which comprised cycling or rowing, followed by dynamic stretching and warm-up exercises using an empty barbell (20 kg), such as good mornings or back squats. Following the warm-up, a load–velocity profile was firstly assessed for each participant, by instructing them to squat three times with maximum lifting velocity with a loading of 45, 55, 65, 75 and 85% of their estimated 1RM. The participants were instructed to pause for 3 min between each load. Secondly, after a pause of 5 min, the actual 1RM of each participant was assessed. Here, the choice of load and pause between attempts was left to the participants to improve their self-confidence before the lift. In case of failure, a second attempt was given, depending on the feedback and fatigue level of the participant. Thirdly, after a pause of 10 min to ensure the best possible recovery, a set-to-exhaustion with 80% of the achieved 1 RM was performed. Participants were explicitly guided to pause momentarily between repetitions, ensuring that a brief but distinct interval was present between repetitions. This measure was implemented to enhance the separability of individual repetitions for more accurate data analysis.

2.3. Data Collection

Velocity parameters were independently but simultaneously measured with an optical 3D motion capture system (as the criterion), two Apple Watch 7s (version: 9.0), both wrist-worn and barbell-mounted, and an Enode Pro device (version: 1.1.6, firmware: 5.2.0) (Figure 2).

Figure 2. Setup of measurement devices. A1: Vicon infrared camera with A2: reflective markers on both proximities of barbell. B1: barbell-mounted Apple Watch, B2: wrist-worn Apple Watch, C1: barbell-mounted Enode Pro, D1: smartwatch not included in study.

The barbell-mounted Apple Watch was positioned adjacent to the left hand of the subject, approximately 60 cm from the left extremity of the barbell. Depending on the participants preferences, the position was adjusted slightly. The wrist-worn Apple Watch was firmly affixed to the left wrist to eliminate any device movement during data acquisition.

The Enode Pro was fixed to the outermost place of the inner part of the barbell. For optical motion capture, two reflective markers were attached on both sides of the barbell, still enabling the change of weight plates between tests, with ten infrared cameras (Vantage 5, Vicon Motion Systems Ltd., Oxford, UK) placed around the participant to record the motion of the markers.

The Enode Pro device was operated via a tablet (iPad Pro, Apple Inc., Cupertino, CA, USA) and calibrated according to instructions of the corresponding app. Following calibration, the device remained attached to the barbell and switched on throughout the entire testing session of each subject. For standardization, it was rebooted and recalibrated after each participant, or in the event of any errors.

To facilitate data collection from the Apple Watch, a Node.js server application equipped with an user interface was developed to stream and receive mobile data via the Sensorlog [26] application (version 5.2). The Node.js application was utilized for all interactions with the Apple Watches during the sessions.

To facilitate communication between the Apple Watch and the paired iPhone (iPhone 13 Pro, Apple Inc., Cupertino, CA, USA), a local network was established, and both iPhones were connected to it during the testing session. Thus, the Apple Watch could communicate with the paired iPhone as its parent device through the Node.js server. To manually calibrate the Apple Watch on the wrist prior to each set, participants were instructed to perform a functional calibration movement that involved swinging the arms at self-selected speed for approximately 30 s. This was not utilized any further in this work.

All Vicon cameras were controlled from an Antec WorkBoy desktop (Antec, Taipei, Taiwan) running Vicon Nexus software (version 2.9, Vicon Motion Systems Ltd., Oxford, UK).

2.4. Data Processing

Velocity parameters from the Vicon optical motion capture and the Apple Watch were derived by processing the motion data using the procedures outlined in the Supplementary Materials (Criterion: Supplementary Material S1, Apple Watch: Supplementary Material S2). Thereby, both markers on either side of the barbell were used to derive the displacement and velocity at 60 cm from the left side of the barbell. The resulting velocity magnitudes from the optical motion capture were used as the validation criterion for comparison with the velocity data from the Apple Watch. To ensure accurate identification of the concentric phases, all repetitions from both modalities were graphically reviewed. Velocity parameters from the Enode Pro were obtained through the export feature of the corresponding iOS application. In our study, the Enode Pro device encountered some recognition errors, in which repetitions were recorded inaccurately, leading to instances of duplicates or neglection. In case of such errors, the entire set was excluded for statistical analysis. Additionally, a validity check was conducted on the velocity parameters, where the criterion Vpeak (peak concentric velocity) > Vmean (mean concentric velocity) was evaluated prior to statistical analysis.

2.5. Statistical Analyses

The validity of the Apple Watch was assessed against the criteria for Vmean, Vpeak and Vprop [27] (propulsive concentric velocity) separately, using the three-tier approach recommended by Hopkins [28]; this approach comprises a Pearson's correlation coefficient (r), a calibration equation, and the standard error of the estimate (SEE). This analysis was carried out for (1) the whole velocity spectrum, (2) repetitions with a concentric criterion duration below 1.25 s, and (3) those with a concentric criterion duration above and equal to 1.25 s, which were categorized as "total", "slow" and "fast", respectively. The time threshold was chosen accordingly, to build two similarly sized groups with two velocity spectra ("slow" and "fast").

For the calibration equation, ordinary least product (OLP) regression was used [29], as both the criterion and Apple Watch measurements were subject to random measurement errors. To assess the effect of the Apple Watch's placement, the OLP between the wrist-worn and barbell-mounted Apple Watch was used. The SEE was calculated from residuals of the OLP calibration equation, according to Siegel [30] and Fritschi et al. [12]. Correlation coefficients were interpreted as large, very large, and extremely large for thresholds of 0.5, 0.7, and 0.9, respectively [31].

According to Fritschi et al. [4], SEE values less than 0.1 m/s and greater than 0.3 m/s are considered low and high-precision values, respectively, whereas SEE values between 0.1 and 0.3 m/s are considered moderately precise. The chosen values of 0.1m/s were adequate for identifying a 30% velocity loss at relatively heavy loads. Proportional measurement bias was considered to exist if the 95% confidence limits of the calibration slope did not include 1, while a fixed measurement bias was considered to exist if the 95% confidence limits of the calibration intercept did not include 0 [29]. This issue is discussed further in the subsequent sections, as the use of an OLP with sufficient high precision allows for the correction of measurement biases. All statistical analysis was performed with Python, employing the SciPy [32] and Pingouin [33] libraries.

3. Results

3.1. Data Set Description

Data from a total of 599 repetitions of the free-weighted back squat in 22 participants were acquired (complete data set in Supplementary S3). After post-processing and elimination of repetitions that exhibited clear errors, the data for validation purposes included 578 (97%), 547 (91%), and 496 (83%) paired observations for the barbell-mounted Apple Watch–Vicon, the wrist-worn Apple Watch–Vicon, and the barbell-mounted Enode Pro–Vicon, respectively (Table 1). The connection between the SensorLog App and the Node.js Server application experienced some timeouts. It was found that this issue was more prevalent for the wrist-worn compared to the barbell-mounted Apple Watch (49 vs. 19 reps). Additionally, the results indicated that the Enode Pro exhibited a higher number of recognition errors compared to the Apple Watch (17 vs. 2–3 reps). Furthermore, 17 repetitions of the Enode Pro failed the validity check, because the Vmean value was larger than the Vpeak velocity.

Table 1. Descriptive summary of recorded repetition data. Data pairs are displayed as the total numbers (n) over all three velocity spectra. A total of 599 repetitions were performed during data collection, of which 19 barbell-mounted and 49 wrist-worn Apple Watch repetitions were missed due to connection issues with the Nodejs server. A total of 70 repetitions were missed by the Enode Pro device due to connection issues, segmentation errors (leading to duplication or neglection of repetitions), or following the validity check (Vpeak > Vmean). The Enode Pro data of one participant were excluded from further analysis because of the large deviation in their squat technique.

Device	Total	Fast	Slow	Connection Issue	Segmentation Error	Validity Check	Ghost Reps	Participant Excluded	Handling Error
AW Barbell	578	290	288	19	2	-	-	-	-
AW Wrist	547	282	265	49	3	-	-	-	-
Enode Pro Barbell	496	258	238	36	17	17	25	1 (27 reps)	6

3.2. Validity

3.2.1. Precision

The indicators of the precision (SEE, correlation coefficient (r)) of all devices are visualized in Figure 3 and in Table 2 (complementary information can be found in Supplementary Material: Tables S1–S3). Additionally, Figure 3 shows the influence of the velocity spectrum on the precision.

The barbell-mounted Apple Watch yielded the highest correlation coefficients (r = 0.879–0.981) in comparison to the Vicon data as criterion, followed by the Apple Watch worn on the wrist (r = 0.774–0.969) and lastly the Enode Pro (r = 0.110–0.981). Particularly for slow velocities, a weak correlation between the Enode Pro and Vicon was found (r = 0.110–0.768). This trend was also confirmed by the SEE values (Figure 3d–f). Here, the values for the barbell-mounted Apple Watch never exceeded 0.1 m/s, closely followed by the wrist-worn Apple Watch, which always had a slightly higher SEE value (0.56–0.114 m/s). The SEE values for Enode Pro were in a similar range for Vmean (0.036–0.078 m/s) and Vprop (0.076–0.114 m/s), but considerably higher for Vpeak (0.180–0.188 m/s).

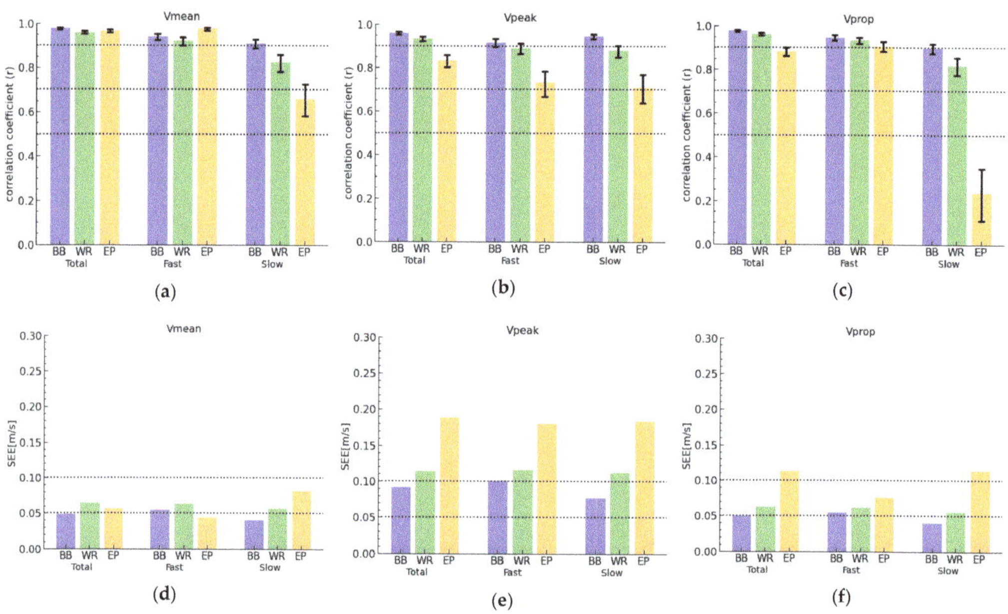

Figure 3. Bar plots indicate the values of the Correlation coefficient (r, top row: **a–c**) and standard error of estimate (SEE, bottom row: **d–f**) of mean, peak, and propulsive velocity measures (i.e., Vmean, Vpeak, Vprop) for all three devices, i.e., the Apple Watch on the barbell (BB), Apple Watch on the wrist (WR) and Enode Pro (EP), in comparison to Vicon as criterion. As described in the methods, the captured repetitions were categorized as total, fast and slow velocity spectra to be separately analyzed. The 95% confidence interval is visualized via the error bars. The dotted line visualize thresholds which were discussed in Section 2.5 Statistical Analyses.

3.2.2. Accuracy

The parameters for the OLP regression are presented in Table 2. OLP results for the different velocity zones can be found in the Supplementary Materials. The slope coefficients demonstrated a consistent pattern for all repetitions across the full velocity spectra and parameters for both Apple Watches (barbell-mounted and wrist-worn). For both devices, the Vpeak showed a larger divergence to the slope value 1 compared to Vmean and Vprop.

Otherwise, the barbell-mounted Enode Pro device displayed shallower slope gradients for Vpeak and Vprop. The different results in accuracy for the Apple Watches compared to Enode Pro are addressed in detail in the discussion section.

Table 2. VBT devices in comparison to Vicon as the criterion: calibration equation parameters with confidence limits (in brackets), standard error of estimate (SEE), and correlation coefficient (r). The slope, intercept and SEE were calculated using OLP regression, as outlined in the method section.

Accuracy					
Device	**Parameter**	**Slope**	**Intercept**	**SEE (m/s, %)**	**Correlation Coefficient (r)**
AW Barbell	Vmean	1.022 [1.004, 1.041]	0.001 [−0.010, 0.012]	0.049 8.1%	0.976 [0.971, 0.979]
	Vpeak	1.112 [1.087, 1.138]	0.023 [−0.006, 0.052]	0.092 7.2%	0.959 [0.952, 0.965]
	Vprop	1.045 [1.027, 1.064]	−0.006 [−0.016, 0.004]	0.05 8.3%	0.977 [0.973, 0.981]
AW Wrist	Vmean	1.016 [0.992, 1.041]	−0.008 [−0.023, 0.007]	0.064 10.4%	0.959 [0.952, 0.965]
	Vpeak	1.105 [1.072, 1.139]	0.041 [0.003, 0.078]	0.114 8.8%	0.934 [0.922, 0.944]
	Vprop	1.029 [1.006, 1.052]	−0.013 [−0.027, −0.0]	0.062 10.4%	0.964 [0.957, 0.969]
Enode Pro Barbell	Vmean	1.065 [1.041, 1.090]	−0.048 [−0.063, −0.033]	0.059 9.6%	0.966 [0.959, 0.971]
	Vpeak	0.967 [0.921, 1.016]	0.185 [0.131, 0.237]	0.188 14.8%	0.833 [0.804, 0.858]
	Vprop	0.990 [0.949, 1.032]	0.038 [0.008, 0.055]	0.113 19.1%	0.882 [0.860, 0.900]

3.3. *Precision of Wrist Placement*

The effects of the placement of the Apple Watch on the precision are presented in Table 3. Precision indicators displayed a high correlation (r = 0.942–0.969) and low SEE values (SEE = 0.06–0.09 m/s) between the wrist-worn and barbell-mounted Apple Watch.

Table 3. Analysis of the effects of placement (wrist vs. barbell) of the Apple Watch. Calibration equation parameters with confidence limits (in brackets), standard error of estimate (SEE), and correlation coefficient (r). The slope, intercept and SEE were generated using OLP regression, as outlined in the method section.

Wrist-Worn vs. Barbell-Mounted Apple Watch		
Parameter	**SEE (m/s, %)**	**Correlation Coefficient (r)**
Vmean	0.063 10.6%	0.958 [0.950, 0.964]
Vpeak	0.09 7.9%	0.951 [0.942, 0.958]
Vprop	0.06 10.4%	0.963 [0.957, 0.969]

4. Discussion

This study examined the validity of two inertial sensors (Apple Watch and Enode Pro) to derive velocity parameters during VBT with free-weighted back squats in healthy, recreational strength training athletes (compared to Vicon as the criterion). To our knowledge, our study is the first to use the IMU sensors of a commercially available smartwatch in a VBT validation study. Furthermore, only a few studies to date have assessed the accuracy of the Enode Pro device across different loads and exercises (including the back squat) compared to the Apple Watch and Vicon as the criterion [12,19,23].

4.1. Importance of Precision

It is generally agreed among practitioners that precision is of greater significance than accuracy when monitoring VBT using a specific device [12]. Here, it is important to note that the present segmentation algorithm developed for the Apple Watch did not incorporate any correction of fixed or proportional biases, and that the present findings are specific to the free-weighted back squat exercise during VBT; this is because different exercises and execution guidelines may yield varying levels of precision. The author suggests that potential developers of VBT applications should consider incorporating bias correction into their algorithms to improve inter-device reliability. However, although some slight over- or underestimations may exist, the precision parameters presented in Figure 3 are considered a good measure for the validity rating in this study.

Some studies have reported a decrease in the precision of IMU devices depending on the velocity of repetitions, particularly in overestimations at higher velocities [16,19]. In our work, we observed a loss in precision using the Enode Pro device compared to Vicon, with an underestimation at lower velocities (Figure 3). This loss in precision at lower velocities was not observed using either Apple Watch. The lower precision of the Enode Pro for slow repetitions could be attributed to the fixation mechanism of the device or to the low velocities assessed in this study. The Enode Pro fixation is solely reliant on a magnetic component as opposed to being tightly strapped to the barbell, as is the case with the barbell-mounted Apple Watch and for other Enode Pro studies [12,19,23,34]. The influence of low velocities on IMU precision might be valuable to assess in further studies.

Unlike the slow repetitions, the mean lifting velocity of the Enode Pro achieved higher precision for fast repetitions compared to the Apple Watch, which supports the quality of the IMU sensor. This finding raises the suspicion that the observed errors for slow propulsive (Vprop) repetitions using the Enode Pro might originate from segmentation inaccuracy on the software side. The propulsive lifting velocity (Vprop), by definition, shares a significant overlap with the mean lifting velocity (Vmean), differing solely in the criterion utilized to define the end of a repetition (Vprop: acceleration below 1g, Vmean: velocity equaling zero). Hence, it is unexpected and unclear why Enode Pro exhibited poorer precision in Vprop in comparison to Vmean, particularly for the slow velocity spectrum, which was not the case for the resulting velocity parameters of the Apple Watch (Figure 3). This observation highlights the significance and impact of the embedded segmentation algorithm on the software side, given that the underlying raw acceleration data are identical for both velocity parameters. Consistent with the literature [11,12,17–19], we found all three devices to be valid for measuring Vmean, with the exception of slow repetitions measured by the Enode Pro device.

The peak lifting velocity of a repetition differs from the mean and propulsive velocities in that it is heavily reliant on the segmentation algorithm and the starting frame when differentiating the acceleration data. Even slight shifts in the segmentation spanning a few time frames may result in substantial errors. This challenge is further compounded when employing a real-time segmentation algorithm, as opposed to post-processing the acceleration data using offline segmentation. Additionally, peak velocity represents a singular snapshot of a time series, whereas mean velocity is calculated as the average across the entire concentric phase of the movement.

4.2. Connectivity and Segmentation

To achieve optimal compliance for various applications such as training interventions, prevention programs, and rehabilitation, it is essential that an athlete experiences as little frustration and exhaustion as possible in their use of technological devices. In the present work, connection issues that resulted in the omission of multiple repetitions were experienced using all devices. The progression of time revealed an increase in connectivity difficulties with the Apple Watch, leading to an increasing number of missed recordings when worn on the wrist. These accumulated connectivity issues were likely caused by the experimental protocol, which necessitated the execution of a calibration movement using the wrist-worn Apple Watch, which then prolonged the recording period. The actual source of the connectivity issue remains, however, unclear, and may have originated from either the Bluetooth connection between the Apple Watch and the iPhone or the network connection between the iPhone and the NodeJs server, respectively. However, the author is confident that the occurrence of such missed repetitions can be reduced with a more refined recording methodology executed by trained professionals.

Furthermore, it is noteworthy that the offline application of the present segmentation algorithm yielded only two (barbell) and three (wrist) inaccurately segmented repetitions, thereby corroborating the robustness of the applied algorithm and the results obtained. For the Enode Pro device, a total of 36 repetitions (accounting for 6.0% of the total) were missed due to connection issues. These results align with previous findings by Jukic et al. (free-weighted back squats, 184 repetitions, 4.6%), Feuerbach et al. (Smith machine back squats, 50 repetitions, 5%), and Fritschi et al. (different free-weighted exercises, 35 repetitions, 5.0%) [12,19,24]. Here, Jukic et al., suggested that Bluetooth-related problems may be the underlying cause of the missed repetitions [24].

Additionally, the presence of a considerable number of "ghost repetitions", (i.e., caused by the re-racking of the barbell) indicates that the Enode Pro has implemented a highly sensitive segmentation algorithm for squat-specific exercise recognition, which may have resulted in inaccuracies, particularly during slow, unsteady and jittery repetitions. Setting a larger threshold for the minimal range of motion (ROM), which was possible in the Enode Pro app for individually created exercises (the exercise "barbell back squat" was chosen in this study), may have prevented these ghost repetitions. The version (1.1.5) of the Enode Pro app has addressed the issue of "ghost repetitions" in the beta version, and further studies should be sure to incorporate these changes. The algorithm developed for this study for the Apple Watches includes a correction for the device's high sensitivity during segmentation, achieved by applying a minimal ROM threshold (0.3 m). Therefore, we hypothesize that by adjusting the minimal ROM, this low precision could be improved.

The extent of measurement errors encountered, particularly with the Enode Pro, was unexpected. Based on the validity outcomes, it is hypothesized that the issue may be attributed to software-related factors, given the challenges associated with the task of segmentation, especially in real time. Our app version 1.1.6 and firmware version 5.2.0 differ from those utilized in other studies [12,19,24]. Furthermore, given the potential software-related nature of the issue, it is crucial to report both the firmware and software versions of the devices utilized for validation purposes, to enable meaningful comparisons between publications.

4.3. *Placement of VBT Device*

Our study demonstrates that both the wrist-worn and barbell-mounted Apple Watch exhibit high precision for Vmean and Vprop parameters across the full velocity spectrum. While the precision is slightly inferior to that reported by Fritschi et al. [12], who analyzed the effect of the attachment point on the barbell, it still falls within an acceptable range (Table 3, SEE < 0.1 m/s, correlation coefficient (r) > 0.9). One explanation for the higher validity of the barbell-mounted Apple Watch compared to the wrist could be the potential accumulation of interference caused by the movement of the arm during the exercise, which is not present when using a rigid barbell. This is supported by the results of the Enode Pro, which are slightly worse compared to the barbell-mounted Apple Watch, and superior to the wrist-worn Apple Watch. However, the utilization of an Apple Watch worn on the wrist has the potential benefit of enhanced user-friendliness, time-efficiency, and cost-effectiveness in VBT. This is particularly relevant given that Thompson et al. [35] identified the significant drivers for coaches to purchase VBT technology as the combination of a good price point and functionality, emphasizing the crucial importance of user-friendliness.

Nonetheless, to achieve higher levels of precision, as may be necessary in the case of coaching professional athletes or for research, we recommend attaching the IMU device directly to the barbell, using a secure fixation mechanism that minimizes motion relative to the barbell.

4.4. Limitations

In the current investigation, each of the three systems, namely the Apple Watch, the Enode Pro device and the Vicon system, were recording data autonomously, and the resulting velocity parameters were derived separately for each data modality. There was no synchronization established among these devices during data recording, which may have introduced some degree of variability into the results. The Enode Pro device, in adherence to standard usage protocols, was secured solely via its magnetic backside, which might have caused some inaccuracies due to device movement during squat performance testing. Due to practical constraints, the positioning of the Enode Pro device was not central, but rather offset from the midpoint of the barbell, contrary to the centrally located Apple Watch (see Figure 2). This divergent positioning might have introduced minor discrepancies in the collected measurement data, as suggested by previous research [12]. Yet, the comparison of the velocity parameters of the Enode Pro device was a secondary objective of the present work, with the primary focus being the assessment of the validity of the Apple Watch for measuring barbell kinematics.

5. Conclusions

The use of the IMU sensor in the Apple Watch as a tool for monitoring movement velocities during free-weighted back squats has shown promising results in terms of its validity. If these results are replicated by other studies, they could lead to the development of an affordable and user-friendly VBT device. However, it is important for practitioners to take into account the impact of the attachment site and the adopted segmentation algorithm on the precision of the resulting velocity parameters.

Our study confirmed previous research on the validity of the Enode Pro device, but also raised questions regarding repetitions at slow lifting velocities, which warrant further investigation. In terms of validity rankings across all 599 repetitions that were analyzed in this study, the barbell-mounted Apple Watch was found to be the most valid for assessing propulsive and peak lifting velocity, followed by the wrist-worn Apple Watch and the barbell-mounted Enode Pro device. For mean lifting velocities, the barbell-mounted Apple Watch led to the most accurate results compared to Vicon as the criterion, with slightly less accurate but similar results obtained for the barbell-mounted Enode Pro and wrist-worn Apple Watch. Furthermore, we emphasize the importance of disclosing the results of both software and hardware versions in scientific reports, as changes in validity may occur between versions.

Supplementary Materials: The following supporting information can be downloaded at: https://www.mdpi.com/article/10.3390/sports11070125/s1, Supplementary Material S1: Detailed description of motion capture data processing steps. Supplementary Material S2: Detailed description of acceleration data processing steps. Supplementary Material S3: Complete data set. Table S1: Detailed results (table) including velocity zones (1.25 s $< t \leq$ 1.25 s) of the barbell-mounted Apple Watch. Table S2: Detailed results (table) including velocity zones (1.25 s $< t \leq$ 1.25 s) of the wrist-worn Apple Watch. Table S3: Detailed results (table) including velocity zones (1.25 s $< t \leq$ 1.25 s) of barbell-mounted Enode Pro device.

Author Contributions: B.A. was responsible for conducting the study, including recruitment of participants, data acquisition and data analysis. K.O. helped with Vicon data acquisition. S.J.F. and S.R.L. were responsible for the study design, and all authors contributed to the discussion of results and writing of the manuscript. All authors have read and agreed to the published version of the manuscript.

Funding: This research was funded by the Swiss National Science Foundation (200021_192289/1).

Institutional Review Board Statement: The study was conducted in accordance with the Declaration of Helsinki, and approved by the Ethics Committee of Berne, Nr: 2021-00403.

Informed Consent Statement: Informed consent was obtained from all subjects involved in the study.

Data Availability Statement: The resulting data set is available under the link for Supplementary Materials.

Acknowledgments: The author would like to express sincere gratitude to the participants for their invaluable time and effort in making this study possible. Their dedication and willingness to contribute to this research are greatly appreciated. The author would also like to extend special thanks to M.V. and P.H. for their generous provision of gym facilities, which were critical to the success of this study. Finally, the author would like to thank C.K. for her assistance with data acquisition.

Conflicts of Interest: The author declares no conflict of interest.

References

1. Beattie, K.; Kenny, I.C.; Lyons, M.; Carson, B.P. The Effect of Strength Training on Performance in Endurance Athletes. *Sport. Med.* **2014**, *44*, 845–865. [CrossRef] [PubMed]
2. Schoenfeld, B.J.; Wilson, J.M.; Lowery, R.P.; Krieger, J.W. Muscular Adaptations in Low- versus High-Load Resistance Training: A Meta-Analysis. *Eur. J. Sport Sci.* **2016**, *16*, 1–10. [CrossRef] [PubMed]
3. Suchomel, T.J.; Nimphius, S.; Stone, M.H. The Importance of Muscular Strength in Athletic Performance. *Sport. Med.* **2016**, *46*, 1419–1449. [CrossRef] [PubMed]
4. Westcott, W.L. Resistance Training Is Medicine: Effects of Strength Training on Health. *Curr. Sport. Med. Rep.* **2012**, *11*, 209–216. [CrossRef] [PubMed]
5. Scott, B.R.; Duthie, G.M.; Thornton, H.R.; Dascombe, B.J. Training Monitoring for Resistance Exercise: Theory and Applications. *Sport. Med.* **2016**, *46*, 687–698. [CrossRef]
6. Weakley, J.; Mann, B.; Banyard, H.; McLaren, S.; Scott, T.; Garcia-Ramos, A. Velocity-Based Training: From Theory to Application. *Strength Cond. J.* **2021**, *43*, 31–49. [CrossRef]
7. Lorenzetti, S.; Lamparter, T.; Lüthy, F. Validity and Reliability of Simple Measurement Device to Assess the Velocity of the Barbell during Squats. *BMC Res. Notes* **2017**, *10*, 707. [CrossRef]
8. Balsalobre-Fernández, C.; Kuzdub, M.; Poveda-Ortiz, P.; Campo-Vecino, J.D. Validity and Reliability of the PUSH Wearable Device to Measure Movement Velocity During the Back Squat Exercise. *J. Strength Cond. Res.* **2016**, *30*, 1968–1974. [CrossRef]
9. Balsalobre-Fernández, C.; Marchante, D.; Muñoz-López, M.; Jiménez, S.L. Validity and Reliability of a Novel IPhone App for the Measurement of Barbell Velocity and 1RM on the Bench-Press Exercise. *J. Sport. Sci.* **2018**, *36*, 64–70. [CrossRef]
10. Banyard, H.G.; Nosaka, K.; Haff, G.G. Reliability and Validity of the Load–Velocity Relationship to Predict the 1RM Back Squat. *J. Strength Cond. Res.* **2017**, *31*, 1897–1904. [CrossRef]
11. Balsalobre-Fernández, C.; Marchante, D.; Baz-Valle, E.; Alonso-Molero, I.; Jiménez, S.L.; Muñóz-López, M. Analysis of Wearable and Smartphone-Based Technologies for the Measurement of Barbell Velocity in Different Resistance Training Exercises. *Front. Physiol.* **2017**, *8*, 649. [CrossRef]
12. Fritschi, R.; Seiler, J.; Gross, M. Validity and Effects of Placement of Velocity-Based Training Devices. *Sports* **2021**, *9*, 123. [CrossRef] [PubMed]
13. Oberhofer, K.; Erni, R.; Sayers, M.; Huber, D.; Lüthy, F.; Lorenzetti, S. Validation of a Smartwatch-Based Workout Analysis Application in Exercise Recognition, Repetition Count and Prediction of 1RM in the Strength Training-Specific Setting. *Sports* **2021**, *9*, 118. [CrossRef]
14. Lorenzetti, S.; Bianco, A.; Stefani, L. The "Journal of Functional Morphology and Kinesiology" Journal Club Series: Highlights on Recent Papers in Athletic Training. *JFMK* **2018**, *3*, 49. [CrossRef] [PubMed]
15. Thompson, W.R. Worldwide Survey of Fitness Trends for 2021. *ACSM Health Fit. J.* **2021**, *25*, 10–19. [CrossRef]
16. Pérez-Castilla, A.; Piepoli, A.; Delgado-García, G.; Garrido-Blanca, G.; García-Ramos, A. Reliability and Concurrent Validity of Seven Commercially Available Devices for the Assessment of Movement Velocity at Different Intensities During the Bench Press. *J. Strength Cond. Res.* **2019**, *33*, 1258–1265. [CrossRef] [PubMed]
17. Clemente, F.M.; Akyildiz, Z.; Pino-Ortega, J.; Rico-González, M. Validity and Reliability of the Inertial Measurement Unit for Barbell Velocity Assessments: A Systematic Review. *Sensors* **2021**, *21*, 2511. [CrossRef] [PubMed]
18. Mitter, B.; Hölbling, D.; Bauer, P.; Stöckl, M.; Baca, A.; Tschan, H. Concurrent Validity of Field-Based Diagnostic Technology Monitoring Movement Velocity in Powerlifting Exercises. *J. Strength Cond. Res.* **2021**, *35*, 2170–2178. [CrossRef]
19. Feuerbacher, J.F.; Jacobs, M.W.; Dragutinovic, B.; Goldmann, J.-P.; Cheng, S.; Schumann, M. Validity and Test-Retest Reliability of the Vmaxpro Sensor for Evaluation of Movement Velocity in the Deep Squat. *J. Strength Cond. Res.* **2022**, *37*, 35–40. [CrossRef]
20. Espinosa, H.G.; Thiel, D.V.; Sorell, M.; Rowlands, D. Can We Trust Inertial and Heart Rate Sensor Data from an APPLE Watch Device? *Proceedings* **2020**, *49*, 128. [CrossRef]
21. Auepanwiriyakul, C.; Waibel, S.; Songa, J.; Bentley, P.; Faisal, A.A. Accuracy and Acceptability of Wearable Motion Tracking for Inpatient Monitoring Using Smartwatches. *Sensors* **2020**, *20*, 7313. [CrossRef]

22. Düking, P.; Fuss, F.K.; Holmberg, H.-C.; Sperlich, B. Recommendations for Assessment of the Reliability, Sensitivity, and Validity of Data Provided by Wearable Sensors Designed for Monitoring Physical Activity. *JMIR Mhealth Uhealth* **2018**, *6*, e102. [CrossRef] [PubMed]
23. Menrad, T.; Edelmann-Nusser, J. Validation of Velocity Measuring Devices in Velocity Based Strength Training. *Int. J. Comput. Sci. Sport* **2021**, *20*, 106–118. [CrossRef]
24. Jukic, I.; King, A.; Sousa, C.A.; Prnjak, K.; McGuigan, M.R. *Velocity-Based Approach to Resistance Training: The Reproducibility and Sensitivity of Commercially Available Velocity Monitoring Technologies*; Auckland University of Technology: Auckland, New Zealand, 2022.
25. González-Badillo, J.J.; Sánchez-Medina, L. Movement Velocity as a Measure of Loading Intensity in Resistance Training. *Int. J. Sports Med.* **2010**, *31*, 347–352. [CrossRef]
26. Bernd, T. Sensor Log; Mobile Application Software, Version 5.2. Available online: http://sensorlog.berndthomas.net/ (accessed on 12 October 2022).
27. Sanchez-Medina, L.; Perez, C.E.; Gonzalez-Badillo, J.J. Importance of the Propulsive Phase in Strength Assessment. *Int. J. Sports Med.* **2010**, *31*, 123–129. [CrossRef] [PubMed]
28. Hopkins, W.G. A New View of Statistics: Measures of Validity . Available online: https://www.sportsci.org/resource/stats/precision.html (accessed on 2 January 2022).
29. Ludbrook, J. Linear Regression Analysis for Comparing Two Measurers or Methods of Measurement: But Which Regression? Linear Regression for Comparing Methods. *Clin. Exp. Pharmacol. Physiol.* **2010**, *37*, 692–699. [CrossRef]
30. Siegel, A.F. Correlation and Regression. In *Practical Business Statistics*; Elsevier: Amsterdam, The Netherlands, 2016; pp. 299–354. ISBN 978-0-12-804250-2.
31. Hopkins, W.G.; Marshall, S.W.; Batterham, A.M.; Hanin, J. Progressive Statistics for Studies in Sports Medicine and Exercise Science. *Med. Sci. Sport. Exerc.* **2009**, *41*, 3–12. [CrossRef] [PubMed]
32. Virtanen, P.; Gommers, R.; Oliphant, T.E.; Haberland, M.; Reddy, T.; Cournapeau, D.; Burovski, E.; Peterson, P.; Weckesser, W.; Bright, J.; et al. SciPy 1.0: Fundamental Algorithms for Scientific Computing in Python. *Nat. Methods* **2020**, *17*, 261–272. [CrossRef]
33. Vallat, R. Pingouin: Statistics in Python. *J. Open Source Softw.* **2018**, *3*, 1026. [CrossRef]
34. Held, S.; Rappelt, L.; Deutsch, J.-P.; Donath, L. Valid and Reliable Barbell Velocity Estimation Using an Inertial Measurement Unit. *IJERPH* **2021**, *18*, 9170. [CrossRef]
35. Thompson, S.W.; Olusoga, P.; Rogerson, D.; Ruddock, A.; Barnes, A. "Is It a Slow Day or a Go Day?": The Perceptions and Applications of Velocity-Based Training within Elite Strength and Conditioning. *Int. J. Sport. Sci. Coach.* **2022**, 1–12. [CrossRef]

sports

MDPI

Article

Impact of the Anatomical Accelerometer Placement on Vertical Jump Performance Characteristics

Damjana V. Cabarkapa *, Dimitrije Cabarkapa, Nicolas M. Philipp and Andrew C. Fry

Jayhawk Athletic Performance Laboratory—Wu Tsai Human Performance Alliance, Department of Health, Sport and Exercise Sciences, University of Kansas, Lawrence, KS 66045, USA

* Correspondence: dcabarkapa98@gmail.com

Abstract: With rapid technological development over recent years, the use of wearable athlete monitoring devices has substantially gained popularity. Thus, the purpose of the present study was to examine the impact of the anatomical placement of an accelerometer on biomechanical characteristics of countermovement vertical jump with and without an arm swing when compared to the force plate as a criterion measure. Seventeen recreationally active individuals (ten males and seven females) volunteered to participate in the present study. Four identical accelerometers sampling at 100 Hz were placed at the following anatomical locations: upper-back (UB), chest (CH), abdomen (AB), and hip (HP). While standing on a uni-axial force plate system sampling at 1000 Hz, each participant completed three non-sequential maximal countermovement vertical jumps with and without an arm swing. All devices recorded the data simultaneously. The following variables of interest were obtained from ground reaction force curves: peak concentric force (PCF), peak landing force (PLF), and vertical jump height (VJH). The findings of the present study reveal that the most appropriate anatomical locations to place the accelerometer device when attempting to estimate PCF, PLF, and VJH during a countermovement vertical jump with no arm swing are CH, AB, and UB, and during a countermovement vertical jump with an arm swing are UB, HP, and UB, respectively. Overall, these findings may help strength and conditioning professionals and sports scientists to select appropriate anatomical locations when using innovative accelerometer technology to monitor vertical jump performance characteristics.

Keywords: sport; wearable technology; testing; assessment; performance; force

check for updates

Citation: Cabarkapa, D.V.; Cabarkapa, D.; Philipp, N.M.; Fry, A.C. Impact of the Anatomical Accelerometer Placement on Vertical Jump Performance Characteristics. *Sports* **2023**, *11*, 92. https://doi.org/10.3390/sports11040092

Academic Editors: Rodrigo Zacca, Flávio Antônio de Souza Castro and Rui Miguel Simões de Azevedo

Received: 7 March 2023
Revised: 18 April 2023
Accepted: 19 April 2023
Published: 21 April 2023

1. Introduction

The vertical jump is one of the most commonly used methods for the assessment of an athlete's lower-body muscular power [1,2]. Given that numerous team sports (e.g., volleyball, basketball, soccer, American football) contain a vertical jump component, many strength and conditioning professionals and sports scientists consider it to be one of the essential factors related to successful athletic performance [3–7].

Over the years, various technologies (e.g., force plates, motion capture systems, accelerometers) have been used for the assessment of vertical jump performance characteristics (e.g., jump height, peak concentric force, impulse). Laboratory-based force plate technology has been considered the criterion measure or "gold standard" testing modality for in-depth analysis of vertical jump performance characteristics [8,9]. However, despite being highly reliable, this technology appears to be unsuitable for field assessments due to its low portability, high costs, and requirement of specialized computer equipment for data collection and analysis. Thus, to help strength and conditioning professionals and sports scientists efficiently administer vertical jump assessments on the field, different portable devices such as contact mats, optical timing systems, and accelerometers have often been implemented [10–14].

A considerable amount of scientific literature has examined the validity and reliability of various accelerometer-based technologies for the assessment of vertical jump performance characteristics and has reported mixed findings [10,11,13–19]. In a recently published study, Cabarkapa et al. [16] found that an innovative accelerometer device (StriveTech) was an acceptable testing modality for the assessment of vertical jump height, despite the tendency to overestimate measurements by approximately 3.1 cm when compared to the laboratory-based force plate system as a criterion measure. Similarly, the Myotest accelerometer was shown to overestimate vertical jump height by 8.0 cm and flight time by 6.4% when compared to the force plate [11,19]. Conversely, Rago et al. [13] indicated that the same device underestimated vertical jump height measurements when compared to a marker-based motion capture system. Moreover, the authors noted that its use is appropriate when measuring contraction time and eccentric duration, while optical measuring technology (Optojump) should be the preferred testing modality when assessing vertical jump height and flight time [13]. On the other hand, Hojka et al. [19] suggested that the accelerometers should not be used for measuring countermovement vertical jump kinetic characteristics due to a significant systemic bias toward underestimating peak force, power, and velocity by 167 N, 843 W, and 0.56 $m \cdot s^{-1}$, respectively [19]. However, when fixed to the barbell, this same device was shown to be a valid and reliable tool for measuring force and power production during a barbell back squat and bench press exercise [18].

Another important factor that needs to be considered when using accelerometer technology is its anatomical placement. Depending on the different manufacturers and their recommendations, the accelerometer can be placed on various body locations (e.g., the scapulae, abdomen, sternum, hip, thigh). Previous research reports have suggested that if the goal is to test performance on sport-specific skills, the accelerometer should be attached to the body segment that is performing a specific movement [20–23]. On the other hand, if the goal is to track a player's position, orientation, velocity, and acceleration on the court/field throughout a practice session and/or game, the accelerometer should be placed on the scapulae [20–23]. However, other anatomical placements (e.g., sternum, wrist, head, abdomen, hip) have been used to examine athletes' performance on sport-specific skills as well as quantify the overall workload [22]. For example, in a recently published study, Cabarkapa et al. [24] examined differences between the abdomen (i.e., 5 cm inferior to the umbilicus) and hip (i.e., 5 cm above the greater trochanter) accelerometer placements for the assessment of vertical jump biomechanical characteristics. The authors discovered that peak concentric and landing force and impulse between the two accelerometer placements were not significantly different, while the accelerometer placed on the hip tended to display lower vertical jump height values than the one placed on the abdomen [24].

With the vertical jump being one of the most commonly used methods for the assessment of an athlete's lower-body muscular power and innovative accelerometer technology being an affordable and user-friendly testing modality, it is of critical importance to understand how they can be simultaneously used to adequately monitor an athlete's performance. Thus, the purpose of the present study was to examine the impact of the anatomical accelerometer placement on the biomechanical characteristics of a countermovement vertical jump with and without an arm swing when compared to the force plate as a criterion measure.

2. Materials and Methods

2.1. Participants

Ten males ($\bar{x} \pm$ SD; age = 23.5 $\pm$ 3.4 years, height = 184.4 $\pm$ 9.9 cm, body mass = 94.7 $\pm$ 16.7 kg) and seven females (age = 23.4 $\pm$ 4.5 years, height = 168.0 $\pm$ 8.6 cm, body mass = 67.1 $\pm$ 10.3 kg) volunteered to participate in this study. All participants were healthy, recreationally active individuals without current and/or previous musculoskeletal injuries that could possibly impair jumping performance. All testing procedures were approved by the University's Institutional Review Board and all participants signed an informed consent form.

2.2. Testing Protocol

Upon arrival at the laboratory, each participant performed a standardized warm-up procedure consisting of a set of dynamic stretching exercises (e.g., high knees, butt kicks, forward lunges, lateral skips, and A-skips). Following the completion of the warm-up, four identical accelerometers (StriveTech, Bothell, WA, USA) sampling at 100 Hz were placed directly on the skin at the following anatomical locations: upper-back (UB; 5 cm below the C7 vertebrae), chest (CH; mid-sternum), abdomen (AB; 5 cm inferior to the umbilicus), and hip (HP; 5 cm above the greater trochanter). A disposable razor was used to remove hair from all anatomical spots where needed, and double-sided tape was used to secure accelerometers to prevent any unnecessary movement that could impact the accuracy of the data collected. Additionally, the AB and HP accelerometers were secured with elastic athletic bands and the rest with elastic bands provided by the manufacturer. To fit a diverse group of participants, the bands were offered in different sizes (i.e., small to extra-large). While standing on a uni-axial force plate system (VALD, Force Decks Max, Brisbane, Australia) sampling at 1000 Hz, each participant completed three non-sequential maximal countermovement vertical jumps with an arm swing (i.e., start with arms placed in the front of the body and elbows bent at a 90-degree angle) and three without an arm swing (i.e., hands on the hips). All four accelerometers recorded the data simultaneously with a force plate. The initiation of the vertical jump was defined as the time point at which the derived ground reaction force decreased by >20 N from the baseline (i.e., system mass), after which peak concentric force and airtime were easily detectable [25,26]. In order to minimize the possible influence of fatigue, each set was separated by a 1–2-minute rest, and each repetition was separated by a 30–60-second rest interval. The overall number of jumps performed across all participants was 126.

2.3. Variables

Ground reaction force curves were computed from raw vertical acceleration data by adding acceleration due to gravity (9.81 $m \cdot s^{-2}$) and multiplying the resultant value with the subject's body mass, from which the following dependent variables were derived: peak concentric force (PCF; highest value observed during the concentric phase of the jumping motion), peak landing force (PLF; highest value observed during the landing phase of the jumping motion), and vertical jump height (VJH; i.e., calculated based on the flight time using the following equation [14,16]; $(t^2 \cdot g)/8$; $g = 9.81\ m\ s^{-2}$, t = time in the air (sec)—determined as a change in the time between the first and last time point when the ground reaction force curve crosses zero value. The same dependent variables were obtained from the force plate system software (VALD, Force Decks Max, Brisbane, Australia). The average value obtained across three jump trials from each accelerometer and the force plate system was used for statistical analysis purposes.

2.4. Statistical Analysis

Descriptive statistics, means and standard deviations ($\overline{x} \pm$ SD) were calculated for each dependent variable. The Shapiro–Wilk test and Q–Q plots corroborated that the assumption of normality was not violated. Independent t-tests were used to examine statistically significant differences in PCF, PLF, and VJH obtained from accelerometers placed at four different anatomical locations (AB, CH, HP, and UB) and a force plate as a criterion measure, for both countermovement vertical jumps with and without an arm swing. Cohen's d was used to calculate the measure of effect size (ES; i.e., d = 0.2 small effect, d = 0.5 moderate effect, d = 0.8 large effect) [25]. Intra-class correlation coefficients (ICC) and Bland–Altman plots based on Giavarina et al. [27] recommendations were used to calculate and graphically represent the absolute agreement between the measurements, respectively. Statistical significance was set a priori to $p < 0.05$. All statistical analyses were performed in SPSS statistical software (Version 28.0; IBM Corp., Armonk, NY, USA).

3. Results

Descriptive statistics, means, and standard deviations ($\bar{x} \pm SD$), are presented in Table 1. Bland–Altman plots for each dependent variable and accelerometer placement in comparison to the force plate as a criterion measurement are presented in Figures 1–6.

No significant differences in PCF for countermovement vertical jumps with no arm swing were found between the force plate and AB, CH, HP, and UB. PLF was significantly lower for CH when compared to a force plate, while no differences were observed for AB, HP, and UB. Also, no significant differences in VJH were found between the force plate and CH, HP, and UB, while the AB displayed significantly greater magnitudes (Table 2).

For countermovement vertical jumps with an arm swing, a significant difference in PCF was found for AB and HP, with both being greater in magnitude when compared to the force plate as a criterion measure, while no differences were observed for CH and UB. AB, CH, and UB all displayed significantly lower PLF when compared to the force plate, while no difference was observed for HP placement. Additionally, no significant differences in VJH were observed between the force plate and AB, CH, HP, and UB (Table 3).

Table 1. Descriptive statistics, means and standard deviations ($\bar{x} \pm SD$), for peak concentric force (PCF), peak landing force (PLF), and vertical jump height (VJH) during countermovement vertical jumps performed with and without an arm swing.

No Arm Swing	PCF [N]	PLF [N]	VJH [cm]
Abdomen	2072.2 ± 698.4	3241.1 ± 902.2	29.3 ± 7.4 *
Chest	1923.9 ± 534.9	2851.7 ± 799.1 *	28.7 ± 7.3
Hip	1984.8 ± 547.0	3162.9 ± 875.5	25.6 ± 7.3
Upper-back	1559.0 ± 586.2	2978.8 ± 948.3	23.9 ± 5.9
Force plate	1848.9 ± 467.3	3621.8 ± 993.6	24.4 ± 6.0
Arm Swing	**PCF [N]**	**PLF [N]**	**VJH [cm]**
Abdomen	2471.2 ± 714.0 *	3092.1 ± 830.1 *	35.2 ± 10.1
Chest	2253.7 ± 736.7	3026.7 ± 847.3 *	33.7 ± 8.3
Hip	2323.5 ± 552.2 *	3224.0 ± 760.1	31.6 ± 8.2
Upper-back	2062.6 ± 717.3	3108.4 ± 932.6 *	28.4 ± 8.2
Force plate	1940.6 ± 468.1	3872.5 ± 1118.4	29.5 ± 7.6

Note: * significantly different when compared to force plate as a criterion measure ($p < 0.05$).

Table 2. Statistical analysis parameters (p-value; intra-class correlation coefficients—ICC; effect size—ES) for four anatomical accelerometer placements during countermovement vertical jumps without an arm swing in comparison to a force plate as a criterion measure.

PCF [N]	*p*-Value	ICC	ES
Abdomen	0.296	0.825	0.376
Chest	0.676	0.968	0.149
Hip	0.456	0.956	0.267
Upper-back	0.132	0.842	0.547
PLF [N]	***p*-Value**	**ICC**	**ES**
Abdomen	0.266	0.641	0.401
Chest	0.022	0.632	0.854
Hip	0.176	0.740	0.490
Upper-back	0.071	0.785	0.662
VJH [cm]	***p*-Value**	**ICC**	**ES**
Abdomen	0.045	0.766	0.727
Chest	0.077	0.877	0.643
Hip	0.595	0.970	0.179
Upper-back	0.865	0.803	0.084

Table 3. Statistical analysis parameters (*p*-value; intra-class correlation coefficients—ICC; effect size—ES) for four anatomical accelerometer placements during countermovement vertical jumps with an arm swing in comparison to a force plate as a criterion measure.

PCF [N]	*p*-Value	ICC	ES
Abdomen	0.019	0.619	0.879
Chest	0.162	0.761	0.507
Hip	0.043	0.761	0.748
Upper-back	0.573	0.883	0.201
PLF [N]	***p*-Value**	**ICC**	**ES**
Abdomen	0.033	0.366	1.316
Chest	0.022	0.457	1.398
Hip	0.065	0.517	1.192
Upper-back	0.044	0.648	1.149
VJH [cm]	***p*-Value**	**ICC**	**ES**
Abdomen	0.079	0.772	0.638
Chest	0.140	0.886	0.528
Hip	0.470	0.926	0.266
Upper-back	0.702	0.906	0.139

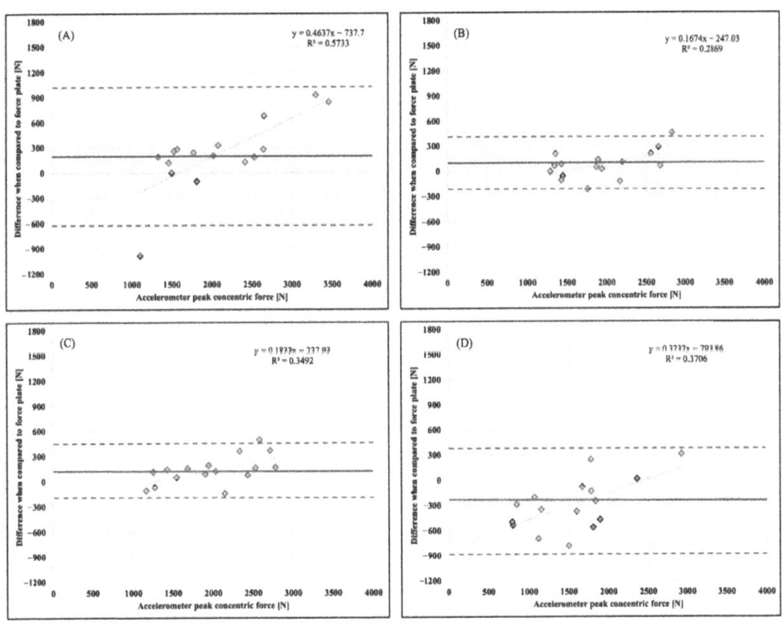

Figure 1. Bland–Altman plot demonstrating measurement agreement in PCF for a CVJ with no arm swing between the force plate and (**A**) abdomen, (**B**) chest, (**C**) hip, and (**D**) upper back accelerometer placement. Solid line—mean difference; dashed line—95% confidence interval; dotted line—linear regression; shaded areas—confidence interval limits for mean and agreement limits.

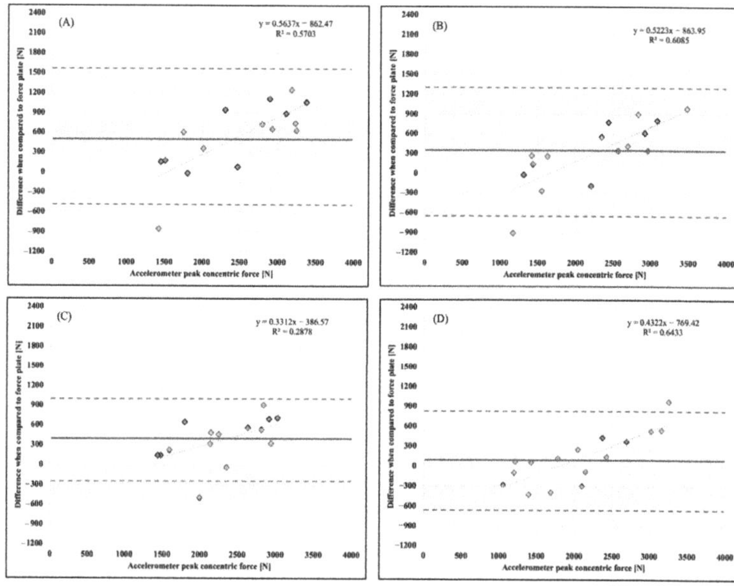

Figure 2. Bland–Altman plot demonstrating measurement agreement in PCF for a CVJ with arm swing between the force plate and (**A**) abdomen, (**B**) chest, (**C**) hip, and (**D**) upper back accelerometer placement. Solid line—mean difference; dashed line—95% confidence interval; dotted line—linear regression; shaded areas—confidence interval limits for mean and agreement limits.

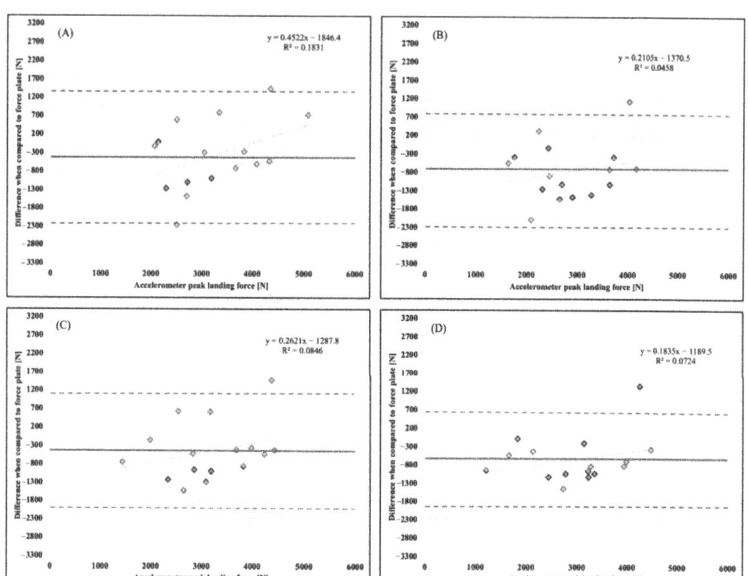

Figure 3. Bland–Altman plot demonstrating measurement agreement in PLF for a CVJ with no arm swing between the force plate and (**A**) abdomen, (**B**) chest, (**C**) hip, and (**D**) upper back accelerometer placement. Solid line—mean difference; dashed line—95% confidence interval; dotted line—linear regression; shaded areas—confidence interval limits for mean and agreement limits.

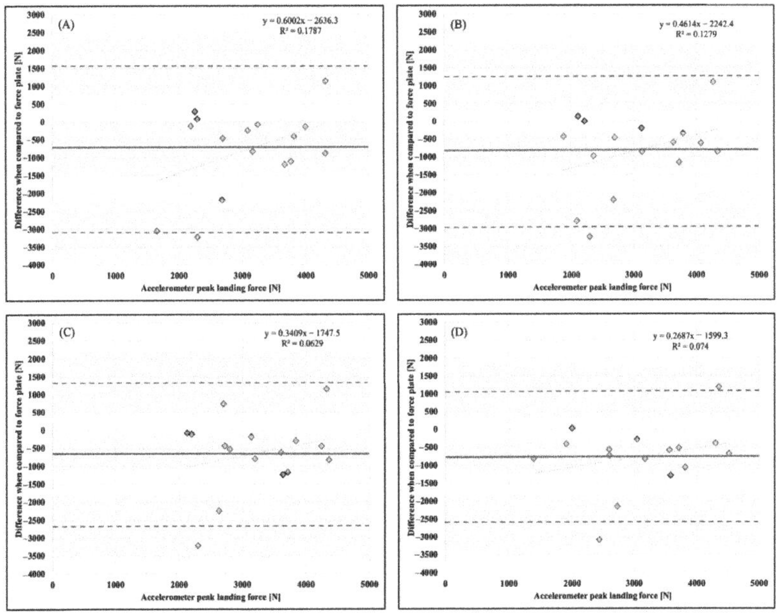

Figure 4. Bland–Altman plot demonstrating measurement agreement in PLF for a CVJ with arm swing between the force plate and (**A**) abdomen, (**B**) chest, (**C**) hip, and (**D**) upper back accelerometer placement. Solid line—mean difference; dashed line—95% confidence interval; dotted line—linear regression; shaded areas—confidence interval limits for mean and agreement limits.

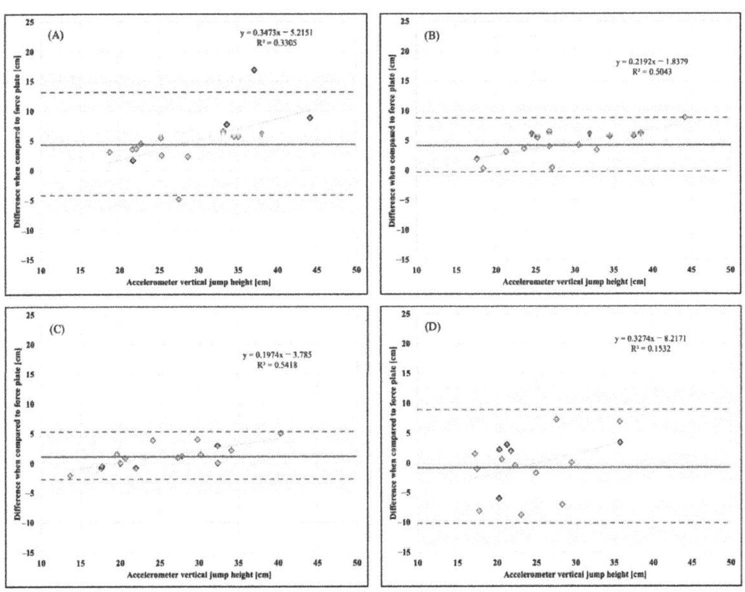

Figure 5. Bland–Altman plot demonstrating measurement agreement in VJH for a CVJ with no arm swing between the force plate and (**A**) abdomen, (**B**) chest, (**C**) hip, and (**D**) upper back accelerometer placement. Solid line—mean difference; dashed line—95% confidence interval; dotted line—linear regression; shaded areas—confidence interval limits for mean and agreement limits.

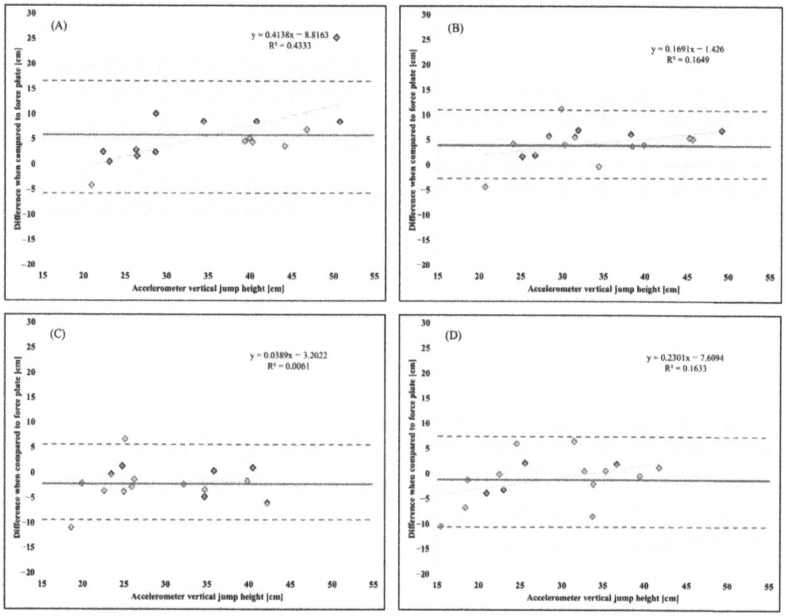

Figure 6. Bland–Altman plot demonstrating measurement agreement in VJH for a CVJ with arm swing between the force plate and (**A**) abdomen, (**B**) chest, (**C**) hip, and (**D**) upper back accelerometer placement. Solid line—mean difference; dashed line—95% confidence interval; dotted line—linear regression; shaded areas—confidence interval limits for mean and agreement limits.

4. Discussion

To the best of our knowledge, this is the first study focused on examining the impact of the anatomical accelerometer placement on biomechanical characteristics (PCF, PLF, and VJH) during a countermovement vertical jump with and without an arm swing when compared to the force plate system as a criterion measure. The findings of the present investigation indicate that AB, CH, HP, and UB accelerometer placements displayed similar PCF values when compared to the force plate system during a countermovement vertical jump without an arm swing (i.e., hands on hips). However, CH and AB accelerometer placements tended to underestimate PLF and overestimate VJH, respectively. On the other hand, when performing a countermovement vertical jump with an arm swing, AB and HP accelerometer placements displayed greater PCF than the values obtained from the force plate system, while no differences were observed for CH and UB anatomical locations. Additionally, despite no statistically significant differences being present in VJH between different accelerometer placements and force plate system, PLF was lower in magnitude for AB, CH, and UB locations.

Previous research has reported mixed findings regarding the use of accelerometer technology for the estimation of PCF during a countermovement vertical jump with no arm swing [19,24,26]. When compared to the force plate as a criterion measure, Hojka et al. [19] found that the Myotest accelerometer placed on the hip tended to underestimate PCF on average by 167 N. In a similar investigation, Howard et al. [26] discovered that the Shimmer accelerometer placed at the same anatomical location overestimated PCF on average by 619 N. The aforementioned findings are contradictory to our results, where no statistically significant differences were observed between each of the four accelerometer placements (AB, CH, HP, and UB) and the force plate as a gold standard testing modality when performing a countermovement jump without an arm swing. However, these results seem to be in agreement with the findings of a recently conducted study that used an identical accelerometer device (StriveTech) and found similar PCF values derived from AB and HP anatomical placements (1753 N and 1776 N) [24]. The observed discrepancies may be attributed to different algorithms and sampling rates used by manufacturers of the previously mentioned accelerometer devices (e.g., 100 Hz vs. 200 Hz). On the other hand, when implementing an arm swing motion during the countermovement vertical jump, a 19–27% increase in PCF was observed for the AB and HP accelerometer placements when compared to the force plate, while CH and UB were similar in magnitude. Although further research is warranted on this topic, it is likely that wearing an accelerometer around the body region with a greater amount of soft tissue (e.g., abdomen) allowed for greater movement of the device, which ultimately resulted in the overestimation of PCF [27]. In addition, it should be noted that the upper-limb contribution to peak ground reaction force has been found to be approximately 31.5% [28]. Thus, this may provide a possible explanation for why the statistically significant differences in PCF have only been detected for AB and HP accelerometer placement when countermovement jumps were performed with an arm swing.

Another important factor to consider when examining countermovement vertical jump performance is PLF. In the present investigation, CH accelerometer placement underestimated PLF during a countermovement vertical jump with no arm swing when compared to the force plate system. However, when the arm swing motion was implemented, significantly lower PLF values were observed across all accelerometer placements (AB, CH, and UB), except for HP anatomical location. This may have occurred due to participants utilizing different landing techniques (e.g., soft landing–knee flexion > 90 deg vs. stiff landing–knee flexion < 90 deg) that have been previously defined by Devita and Skelly [29]. During softer landings, participants tend to attain greater hip and knee flexion and finish with the torso in a less erect position (e.g., lean forward) [29]. This biomechanical alteration may have allowed for greater shock absorption in contact with the ground, especially for the accelerometer placed within the upper-body region (CH and UB), ultimately resulting in underestimation of PLF during both countermovement vertical jumps with and without an arm swing [29]. Moreover, greater discrepancies observed in the arm swing condition could be attributed to the additional movement that occurred when participants added an upper-body contribution (e.g., greater PCF, PLF, and VJH). In addition, it should be noted that currently there is a considerable gap in the scientific literature focused on examining the application of accelerometer technology for quantifying PLF. In addition to the previously mentioned impact of body composition, this is another area that warrants further investigation.

This investigation also examined the differences in VJH between each accelerometer placement (AB, CH, HP, and UB) in comparison to the force plate system as a criterion measure. The AB accelerometer placement significantly overestimated VJH during countermovement vertical jump with no arm swing, while no differences were observed for the remaining anatomical locations (CH, HP, and UB). Similar observations were made by Cabarkapa et al. [16] who found that the accelerometer device placed 3 cm above the umbilicus (i.e., anterior abdomen) tended to overestimate VJH by approximately 3.1 cm during a countermovement vertical jump with no arm swing when compared to the force plate as a gold standard testing modality. The observed similarities may be attributed to researchers using the same accelerometer technology (StriveTech) for VJH assessment as well as similar anatomical placement (i.e., 5 cm vs. 3 cm inferior to the umbilicus). On the other hand, when positioned on HP, Hojka et al. [19] indicated that the Myotest accelerometer tended to overestimate VJH on average by 8 cm, while no significant differences were observed in the present investigation. This may imply that using an HP placement with the accelerometer device used in the present study may yield more accurate VJH measures than the one used by Hojka et al. [19]. In addition, when the countermovement vertical jump was performed with an arm swing, no significant differences in VJH were observed, regardless of the accelerometer anatomical location (AB, CH, HP, and UB). These findings further solidify the applicability of the accelerometer device used by the authors (StriveTech) as an accurate testing modality for the assessment of VJH.

While these findings offer additional insight into the impact of the anatomical accelerometer placement on the assessment of PCF, PLF, and VJH, this study is not without limitations. As previously indicated, body composition was not evaluated, and may have contributed to an excessive accelerometer movement that could increase the margin of measurement error. Additionally, considering that the participants examined in the present study were recreationally active individuals, future research needs to examine if our findings remain identical within a cohort of elite individual and team-sport athletes (e.g., basketball and volleyball players).

In conclusion, based on the smallest measurement error and the greatest level of agreement in comparison to a force plate as a criterion measure, the findings of the present study reveal that the most appropriate anatomical location to place the accelerometer device when attempting to estimate PCF, PLF, and VJH during a countermovement vertical jump with no arm swing are CH, AB, and UB, and during a countermovement vertical jump with an arm swing are UB, HP, and UB, respectively. Overall, these findings may help strength and conditioning professionals and sports scientists to select appropriate anatomical locations when using innovative accelerometer technologies to monitor vertical jump performance characteristics.

Author Contributions: Conceptualization, D.V.C. and D.C.; methodology, D.V.C. and D.C.; investigation, D.V.C., D.C., and N.M.P.; data curation, D.V.C., N.M.P., and D.C.; writing—original draft preparation, D.V.C. and D.C.; writing—review and editing, D.V.C., D.C., N.M.P., and A.C.F. supervision, A.C.F. All authors have read and agreed to the published version of the manuscript.

Funding: This research received no external funding.

Institutional Review Board Statement: The study was conducted according to the guidelines of the Declaration of Helsinki and approved by the Institutional Review Board of the University of Kansas—#00148260.

Informed Consent Statement: Written informed consent was obtained from all subjects involved in the study.

Data Availability Statement: The data presented in this study are available upon request from the corresponding authors.

Acknowledgments: The authors would like to thank participants for their time and willingness to participate in the present investigation.

Conflicts of Interest: The authors declare no conflict of interest.

References

1. Liebermann, D.G.; Katz, L. On the assessment of lower-limb muscular power capability. *Isokinet. Exerc. Sci.* **2003**, *11*, 87–94. [CrossRef]
2. Markovic, G.; Dizdar, D.; Jukic, I.; Cardinale, M. Reliability and factorial validity of squat and countermovement jump tests. *J. Strength Cond. Res.* **2004**, *18*, 551–555. [PubMed]
3. Fry, A.C.; Kreamer, W.K. Physical performance characteristics of American collegiate football players. *J. Strength Cond. Res.* **1991**, *5*, 126–138.
4. Kellis, S.E.; Tsitskaris, G.K.; Nikopoulou, M.D.; Mousikou, K.C. The evaluation of jumping ability of male and female basketball players according to their chronological age and major leagues. *J. Strength Cond. Res.* **1999**, *13*, 40–46.
5. Sheppard, J.M.; Cronin, J.B.; Gabbett, T.J.; McGuigan, M.R.; Etxebarria, N.; Newton, R.U. Relative importance of strength, power, and anthropometric measures to jump performance of elite volleyball players. *J. Strength Cond. Res.* **2008**, *22*, 758–765. [CrossRef]
6. Stolen, T.; Chamari, K.; Castagna, C.; Wisloff, U. Physiology of soccer. *Sport. Med.* **2005**, *35*, 501–536. [CrossRef] [PubMed]
7. Ziv, G.; Lidor, R. Vertical jump in female and male basketball players-A review of observational and experimental studies. *J. Sci. Med. Sport* **2010**, *13*, 332–339. [CrossRef] [PubMed]
8. Aragón, L.F. Evaluation of four vertical jump tests: Methodology, reliability, validity, and accuracy. *Meas. Phys. Educ. Exerc. Sci.* **2000**, *4*, 215–228. [CrossRef]
9. Linthorne, N.P. Analysis of standing vertical jumps using a force platform. *Am. J. Phys.* **2001**, *69*, 1198–1204. [CrossRef]
10. Buckthorpe, M.; Morris, J.; Folland, J.P. Validity of vertical jump measurement devices. *J. Sports Sci.* **2012**, *30*, 63–69. [CrossRef]

11. Castagna, C.; Ganzetti, M.; Ditroilo, M.; Giovannelli, M.; Rocchetti, A.; Manzi, V. Concurrent validity of vertical jump performance assessment systems. *J. Strength Cond. Res.* **2013**, *27*, 761–768. [CrossRef]
12. Glatthorn, J.F.; Gouge, S.; Nussbaumer, S.; Stauffacher, S.; Impellizzeri, F.M.; Maffiuletti, N.A. Validity and reliability of Optojump photoelectric cells for estimating vertical jump height. *J. Strength Cond. Res.* **2011**, *25*, 556–560. [CrossRef] [PubMed]
13. Rago, V.; Brito, J.; Figueiredo, P.; Carvalho, T.; Fernandes, T.; Fonseca, P.; Rebelo, A. Countermovement jump analysis using different portable devices: Implications for field testing. *Sports* **2018**, *6*, 91. [CrossRef] [PubMed]
14. Whitmer, T.D.; Fry, A.C.; Forsythe, C.M.; Andre, M.J.; Lane, M.T.; Huda, A.; Honnold, D.E. Accuracy of a vertical jump contact mat for determining jump height and flight time. *J. Strength Cond. Res.* **2015**, *29*, 877–881. [CrossRef] [PubMed]
15. Bubanj, S.; Stankovic, R.; Bubanj, R.; Bojic, I.; Djindjic, B.; Dimic, A. Reliability of Myotest tested by a countermovement jump. *Acta Kinesiol.* **2010**, *4*, 46–48.
16. Cabarkapa, D.; Fry, A.C.; Hermes, M.J. Accuracy of an experimental accelerometer for assessing countermovement vertical jump height. *Sports Innov. J.* **2021**, *2*, 45–55. [CrossRef]
17. Casartelli, N.; Müller, R.; Maffiuletti, N.A. Validity and reliability of the Myotest accelerometric system for the assessment of vertical jump height. *J. Strength Cond. Res.* **2010**, *24*, 3186–3193. [CrossRef]
18. Comstock, B.A.; Solomon-Hill, G.; Flanagan, S.D.; Earp, J.E.; Luk, H.Y.; Dobbins, K.A.; Dunn-Lewis, C.; Fragala, M.S.; Ho, J.Y.; Hatfield, D.L.; et al. Validity of the Myotest in measuring force and power production in the squat and bench press. *J. Strength Cond. Res.* **2011**, *25*, 2293–2297. [CrossRef]
19. Hojka, V.; Tufano, J.J.; Maly, T.; Stastny, P.; Jebavy, R.; Feher, J.; Zahalka, F.; Gryc, T. Concurrent validity of Myotest for assessing explosive strength indicators in countermovement jump. *Acta Gymnica* **2018**, *48*, 95–102. [CrossRef]
20. Bastida-Castillo, A.; Gómez-Carmona, C.D.; De La Cruz Sánchez, E.; Pino-Ortega, J. Comparing accuracy between global positioning systems and ultra-wideband position tracking systems used for tactical analyses in soccer. *Eur. J. Sport Sci.* **2019**, *19*, 1157–1165. [CrossRef]
21. Bastida-Castillo, A.; Gómez-Carmona, C.D.; De la Cruz-Sánchez, E.; Reche-Royo, X.; Ibáñez, S.J.; Pino Ortega, J. Accuracy and inter-unit reliability of ultra-wideband tracking system in indoor exercise. *Appl. Sci.* **2019**, *9*, 939. [CrossRef]
22. Gómez-Carmona, C.D.; Bastida-Castillo, A.; Ibáñez, S.J.; Pino-Ortega, J. Accelerometry as a method for external workload monitoring in invasion team sports. A systematic review. *PLoS ONE* **2020**, *15*, e0236643. [CrossRef]
23. Nedergaard, N.J.; Robinson, M.A.; Eusterwiemann, E.; Drust, B.; Lisboa, P.J.; Vanrenterghem, J. The relationship between whole-body external loading and body-worn accelerometry during team-sport movements. *Int. J. Sports Physiol. Perform.* **2017**, *12*, 18–26. [CrossRef] [PubMed]
24. Cabarkapa, D.V.; Fry, A.C.; Cabarkapa, D.; Parra, M.E.; Hermes, M.J. Impact of accelerometer placement on assessment of vertical jump performance parameters. *J. Phys. Educ. Sport* **2022**, *22*, 1340–1345.
25. Cohen, J. *Statistical Power Analysis for Behavioral Sciences*, 2nd ed.; Routledge: New York, NY, USA, 1988.
26. Howard, R.; Conway, R.; Harrison, A.J. Estimation of force during vertical jumps using body fixed accelerometers. In Proceedings of the 25th IET Irish Signals & Systems Conference, Limerick, Ireland, 26–27 June 2013; Institution of Engineering and Technology: Stevenage, UK, 2014; pp. 102–107.
27. Giavarina, D. Understanding Bland Altman analysis. *Biochem. Med.* **2015**, *25*, 141–151. [CrossRef]
28. Mosier, E.M.; Fry, A.C.; Lane, M.T. Kinetic contributions of the upper limbs during counter-movement vertical jumps with and without arm swing. *J. Strength Cond. Res.* **2019**, *33*, 2066–2073. [CrossRef] [PubMed]
29. Devita, P.; Skelly, W.A. Effect of landing stiffness on joint kinetics and energetics in the lower extremity. *Med. Sci. Sports Exerc.* **1992**, *24*, 108–115. [CrossRef]

sports

MDPI

Review

The eSports Medicine: Pre-Participation Screening and Injuries Management—An Update

Agostino Rossoni [1], Marco Vecchiato [2], Erica Brugin [3], Eliana Tranchita [4,5], Paolo Emilio Adami [6,7], Manuela Bartesaghi [1], Elena Cavarretta [8,9] and Stefano Palermi [10,*]

1 Laboratory of Clinical Physiology and Sports Medicine, Department of Experimental Medicine, University of Milano-Bicocca, 20900 Monza, Italy
2 Sports and Exercise Medicine Division, Department of Medicine, University Hospital of Padova, 3651 Padova, Italy
3 Cardiovascular Rehabilitation and Sports Medicine Service, 30033 Noale, Italy
4 Department of Movement, Human and Health Sciences, University of Rome "Foro Italico", 00135 Rome, Italy
5 Department of Cardiac Surgery, Cardiology and Heart Lung Transplant, Bambino Gesù Children's Hospital, IRCCS, 00165 Rome, Italy
6 Health and Science Department, World Athletics, 98007 Monaco, Monaco
7 Laboratoire Motricité Humaine Expertise Sport Santé (LAMHESS), Université Côte d'Azur, 06000 Nice, France
8 Department of Medical-Surgical Sciences and Biotechnologies, Sapienza University of Rome, 04100 Latina, Italy
9 Mediterranea Cardiocentro, 80122 Napoli, Italy
10 Public Health Department, University of Naples Federico II, Via Pansini 5, 80131 Naples, Italy
* Correspondence: stefanopalermi8@gmail.com

Abstract: Recently, electronic sports (eSports) became one of the growing forms of new media due to the wide diffusion of games and online technologies. Even if there is still a debate about the definition and characterization of eSports, eAthletes train heavily, compete in tournaments, must abide by competition, association, and governing body rules, just like all other athletes. Furthermore, as in any other competitive discipline, there can be injuries. Aberrant sitting posture, repetitive movements, screen vision, prolonged playing hours, and a sedentary lifestyle can lead to several medical hazards in musculoskeletal, ophthalmology, neurological, and metabolic systems. Moreover, several cardiovascular changes occur in eAthletes. This paper aims to explore the different injuries that can occur in a professional eAthlete, suggesting how every high-level gamer could benefit from a pre-participation evaluation and a correct injury prevention strategy.

Keywords: electronic sport; eSports; eGames; eAthletes; sports medicine; cardiovascular prevention; injury prevention

check for updates

Citation: Rossoni, A.; Vecchiato, M.; Brugin, E.; Tranchita, E.; Adami, P.E.; Bartesaghi, M.; Cavarretta, E.; Palermi, S. The eSports Medicine: Pre-Participation Screening and Injuries Management—An Update. *Sports* **2023**, *11*, 34. https://doi.org/10.3390/sports11020034

Academic Editor: Rodrigo Zacca

Received: 21 December 2022
Revised: 30 January 2023
Accepted: 30 January 2023
Published: 1 February 2023

1. Introduction

Even if a recognized definition is still missing, electronic sport (eSport) has been traditionally described as "a form of sport where the primary aspects of the sport are facilitated by electronic systems, and the input of players and teams are mediated by human-computer interfaces" [1].

During recent years, especially during the COVID-19 pandemic period, eSports became rapidly a growing form of new media thanks to the increasing availability of online games and broadcasting technologies. Some events and tournaments culminate in championships at the regional and international levels, in which players compete against each other, as single players or teams [2–4]. The phenomenon of eSport attracts many spectators: nowadays, people regularly consume eSports events through online-developed streaming platforms [5] (YouTube, Twitch, or Smashcast), gaining a global audience of about half of a billion fans and passing the one billion dollars mark [6].

ESport, as a sport discipline, is growing in recognition worldwide. In China, it has been already defined as a sports activity since the beginning of the 2000s and will be an official competitive sports event in the 19th Asian Games in Hangzhou 2023 [7]; in the USA, eAthletes have been considered professional players since 2013. The phenomenon is so widespread that the International Olympic Committee (IOC) accepted eSports as a sport in 2017 and there have been speculations about their introduction in the Paris 2024 Olympic Games [3,8]. Moreover, in 2021, IOC created the Olympic Virtual Series, composed by baseball, cycling, rowing and motor racing [9].

Even if e-sport has received a wide international acknowledgment, there is still resistance as to whether it can truly be considered as a sport discipline played by real athletes, although it is undoubtful that this discipline has a big impact on players' health. Indeed, several reports have been published [10,11] and this is a growing topic in the scientific literature.

Therefore, this paper aims to review the numerous different injuries that may occur in these athletes, named eAthletes, trying to highlight the potential benefit of a pre-participation evaluation and a correct injury prevention program.

2. Can We Consider ESport as a Physical Sport Discipline?

Before eSport could be classified as a professional sport, two main criteria related to the concept of sports should be explored [12]. The first one is about physical performance and the skillful use of the body or a part of it, similar to what happens in some sports (i.e., in shooting only arms are used). For the second criterion, eSports require institutional stability, with rules for regulation and governance. Even if the different types of eSports make it difficult to achieve universal institutional stability, global eSports organizations already exist, such as the International eSports Federation [13]

There still is no general agreement about the classification of eSports, but they can be differentiated into specific genres such as multiplayer online battlefield arena (MOBA), first-person shooters (FPS), real-time strategy (RTS), or sports simulations [5]. Others classified them as shooter, band simulation, dance simulation, and fitness categories [7]. In every genre, different digital games can be found, and played through various platforms (e.g., computers, consoles, mobile, or virtual reality), with different mechanics and competition rules [14]. All these different modalities of play lead obviously to different types and severity of health problems.

As for other types of sports, we can find professional eSports players and amateur players. Professional players in eSports share many similarities with their traditional physical athlete counterparts: they train rigorously, compete in tournaments, and must abide by competition, association, and governing body rules [15]. They have the same mental and physical issues as physical athletes, sharing some concerns such as stress before competitions or short career life. The eAthlete signs a contract with a professional e-team, receiving salaries and sponsorship; moreover, the infrastructure of eSports can be compared to those of traditional physical team sports [6].

Nowadays professional clubs from different sports such as football, basketball, and F1 racing are increasingly getting involved in eSports to extend their recognition, setting up their e-teams to compete for national and international titles. Scientific literature described several benefits of eSports [7,16,17]:

- No gender differences
- Improvement of the health and well-being of sedentary individuals
- Group participation and competition regardless of location
- Promotion of physical activity for individuals with physical and cognitive disabilities
- Enhancement of cognitive development and school achievement
- Enhancement of social interaction
- Improvement of fine motor skills and visual processing
- Promotion of an effective rehabilitation.

However, some harmful effects are still described [7,16,17]:

- Potential treatments to child safety
- Inappropriate content
- Exposure to violence
- Bullying
- Internet addiction
- Reduction of moderate/vigorous physical activity
- Displacement from active social life and interactions
- Exposure to junk food advertising
- Sleep deprivation
- Vision complaints
- Musculoskeletal overuse.

To achieve top performance in an eSports game, professional eAthletes train on average 5 h per day to improve specific game skills and motor abilities, such as gamepad and keyboard handling, eye-hand coordination, reaction time [18], game knowledge as well as strategy and tactics; during competition, they must be able to sustain high levels of attention and take important decisions under time pressure [6].

3. ESports Health Problems

As for training, medical problems of eAthletes are still scarcely reported in the scientific literature [19] and a systematic analysis of it is still missing in current literature. Nevertheless, scientific research suggests that gaming may have several undesirable physical and mental health effects (Table 1) [11,20]; in the study by Di Francisco-Donoghue et al. [4] eAthletes reported that about half of the players suffered from eye fatigue, neck and back pain, and wrist and hand pain. Therefore, prevention, diagnosis, and treatment of eSports-related issues are needed.

Table 1. Health problems of eSports.

Body Part	Mechanism	Injury
Musculoskeletal system	Aberrant sitting prolonged posture	Neck pain
		"Text neck" syndrome
		Low-back pain
		Radiculopathy
		Gluteal pain
		Hamstring tightness
	Repetitive movements	Shoulder overuse tendinopathy
		Elbow overuse tendinopathy
		Wrist overuse tendinopathy
		Cubital tunnel syndrome
		Carpal tunnel syndrome
		"Gamer's thumb"
		Microtrauma
Ocular system	Long-lasting screen vision	Eye fatigue
		Dry eyes
		Temporary alterations of visual functions in children
	Virtual reality headset gaming	Alteration in the ocular surface and tear film
	Vergence-accommodation conflict	Decreasing vergence ability and accommodative facility
	Increased neural plasticity	Increased visual attentional resources in multi-object tracking

Table 1. *Cont.*

Body Part	Mechanism	Injury
Nutrition	Overeating	Weight increase, obesity
	Energy expenditure, metabolism	Sedentary lifestyle, higher body fat percentage
	Wrong alimentation	Nutrients' deficiency
	Increasing use of caffeine and energy drinks	Cardiac arrhythmias
Neurological system	Long-lasting screen vision	Headache
		Stress
		Insomnia
	Virtual reality headset gaming	"Cybersickness"
Mental Health	Addiction, isolation, several playing hours per game	Gambling
		Depression
		Stress
		Other psychosocial problems

3.1. Musculoskeletal System

To achieve top performance, eAthletes spend several hours per day with repetitive movements using a gamepad, mouse, and keyboard and sitting for an extended period often with sub-optimal posture: these are the two major mechanisms causing musculoskeletal complaints in this sport [21]. Indeed, different platforms of play have important consequences on gamers' posture [22,23]. Neck and back pain are the two most common problems, even if other body districts may be interested [4].

The typical forward displacement of the head while playing leads to anomalies in the cervical spine and muscles, possibly causing repercussions also to the lower back: early degeneration in neural disks and roots, peripheral compressions, and radiculopathies are frequent in eAthletes [21,24]. Appropriate interventions for eAthletes with back pain include exercise therapy, stretching and manual manipulation, together with physical therapies [25], to reduce muscle tension and improve mobility and flexion of the spine [26]. Stretching and strengthening should focus on muscles associated with postural control [27]. Mezieres and McKenzie are two used methods for it.

Competitive eAthletes engage in fine motor movements, with both isotonic and isometric fine contractions: they execute more than 300 high-task movements per minute, predisposing to repetitive and chronic injuries [28,29]. Hand and wrist pain issues are reported by 30% of players [28,29]. Carpal or ulnar tunnel syndromes, De Quervain's tendinopathy, and elbow pain are common diagnoses.

Therefore, a musculoskeletal screening examination is mandatory for eAthletes, also using methods already described [30]. The use of diagnostic instruments is helpful, such as the X scan for cervical spine alignment, magnetic resonance imaging for lumbar herniation, electromyography for radiculopathy [31], and ultrasound for muscle, ligament, and tendon injuries [25].

There are several ergonomics assessment tools available, developed by organizations such as the U.S. National Institute of Occupational Safety and Health, that can be used in screening musculoskeletal and ergonomic complaints in these athletes. Monitor, chair, desktop, keyboard-mouse-controller, and lightning should be the main ergonomics to assess.

3.2. Ophthalmology System

ESports also require visual and attentional stamina [32]. Recently, vision issues likened to the use of digital screens have become a huge public health problem. "Digital eye strain" is a newborn term created to include all aspects of eyesight issues related to long working sessions in front of a digital screen [33]. EAthletes require to have their eyes fixed on a computer screen with excessive exposure to light-emitting diodes [8], and this can impact the natural circadian rhythm [34], especially during tournaments, which might last hours. Moreover, screen time before sleeping has been linked with "screen insomnia," a common among gamers [35]. Recently it has been demonstrated that excessive exposure to blue light spectrum can create retinal and photoreceptor problems [36].

Limiting playtime or using particular devices (e.g., screen glasses) to avoid these hazards should be the main aim in the medical management of these athletes. The American Optometric Association suggests the 20/20/20 rule, taking 20 s of break every 20 min to view something 20 feet away (i.e., take a 20-s break to view something 20 feet away every 20 min) [37].

Assess the visual acuity, evaluate position and alignment of the eyes, and assess extraocular movements, perform the Vestibular Ocular Motor Screen, test saccades and assess convergence and accommodation amplitude should be incorporated into the evaluation of these athletes. Oculomotor performance training can optimize performance [7]: eye exercises are a key component of this strategy.

3.3. Nutrition

Athletes, to compete at their optimum level, need to pay attention to their nutrition; according to the sports discipline, they may require specific calories, macronutrients, and micronutrients to maintain strength and energy and ensure adequate hydration. However, it has been recently demonstrated how eAthletes often have bad nutritional habits [38]: their meals are irregularly consumed, snacking is frequent, and sugared and fried foods are commonly eaten. These unbalanced and largely unhealthy dietary habits raise several health concerns.

A balanced diet is crucial for eAthletes, mainly belonging to a young age group (between 18–25 years) [4], who should need an intake from 1800 to 3000 Kcal/day, according to the World Health Organization [39]. Avoiding a sedentary lifestyle and following a personalized diet should be two main aims for the nutritional management of eAthletes [35]. Body mass index should be routinely evaluated for these athletes, other than nutrition habits using validated questionnaires [40,41]. The evaluation of fat mass and fat-free mass through bioelectrical impendence analysis or plicometry could also be useful.

Inadequate intake of nutrients leads to deficiencies, possible causes of many diseases [42]: it is the cause of deficits of vitamin B12, folate, and zinc that may cause symptoms of fatigue, depression, and dementia. Furthermore, there is growing evidence that specific dietary supplements may enhance cognitive performance, such as caffeine, creatine, or polyphenols [43].

Energy drinks (EDs) are more commonly used by the young population as a source of dietary supplements, especially among eAthletes who spend many hours sitting, sometimes without a proper diet. The common ingredients of EDs are caffeine, sugar, taurine, and ginseng [44]. Caffeine has been classified as a supplement with benefits also in terms of sports performance. Lower doses may improve concentration and reaction times, while higher doses can influence cognitive and physical performance, also in eAthletes [43]. However, it has been demonstrated that caffeine can have pro-arrhythmic and blood pressure-increasing effects, especially when the dosage exceeds 300 mg per day or in susceptible individuals at much lower dosage [45]; other components of EDs can also contribute to cardiovascular side effects, ranging from benign arrhythmias to potentially fatal cardiovascular diseases. Furthermore, the sugar component of EDs induces insulin resistance, and stimulates proinflammatory activity causing dyslipidemia and weight gain [44]. Moreover, although caffeine is currently not on the World Anti-Doping Agency (WADA) List of Prohibited Substances, it is constantly monitored for its possible re-introduction.

Consequently, closer attention should be paid to eAthletes' nutrition and supplementation, such as the use/abuse of some type of EDs, raising awareness about the positive and negative effects of these products. The ASSIST is a helpful screening questionnaire for monitoring players' supplement use [46]. For eSports to be fully acknowledged and accepted among other sports disciplines, it will be necessary that doping control rules and regulations apply similarly as they do in other sports.

3.4. Neurological System and Mental Health

The effects of gaming on the human brain have been recently questioned in literature [47]. Insomnia or sleep deprivation, headache, and stress [48] are frequently found in eAthletes. Long hours of online gaming could be linked to the development of some mental issues such as depression, social phobia, obsession-compulsion disorder, and gaming addiction [48]. Indeed, despite differences in socio-demographic and playing variables exist, eSports and physical sports players tend to exhibit lots of similarities in psychological domains: therefore eSports psychology must not be underestimated, given the recognition of gaming disorder in ICD11 [49], and adequate mental health strategies are needed [50]. The use of some validated questionnaires can be useful to diagnose these problems [7]: the Patient Health Questionnaire (PHQ), followed by the PHQ-9 can screen for major depressive disorders; Generalized Anxiety Disorder (GAD) can be screened using the GAD-2 and GAD-7 questionnaires; Internet Gaming Disorder can be screened using the Internet Gaming Disorder Scale 9-Short Form. Cognitive training, relaxation exercise, and breathing techniques are crucial component of a correct mental injury management, in order to build strong mental skills [7].

Screens of digital devices may interfere with sleep: indeed, blue light suppresses melatonin production, affecting circadian rhythms and impairing cognitive performance [7]. This is a big problem since many eAthletes play during nighttime due to the longer uninterrupted duration [51]. The use of blue light-blocking glasses and a sleep hygiene plan can improve sleep duration and quality [36]. The Pittsburg Sleep Quality Index is a validated questionnaire that could be useful in sleep monitoring [52].

When using virtual reality devices, players may develop symptoms similar to motion sickness, the so-called cybersickness: the human brain must integrate real-time vision, hearing, vestibular and proprioceptive inputs, therefore resulting in a high-task role [53]. On the other side, however, there is evidence indicating how active video games have a positive psychological impact, concerning liking and enjoyment, attitudes, and self-efficacy [54]. Online social interaction, via eGames, has been found to strengthen meaningful feelings and self-regulation, improving some social needs otherwise. Moreover, gaming was shown to be effective also as a therapy against mental disorders, such as attention deficit hyperactivity disorder (ADHD) [55–58]. Recent interesting papers showed how playing Minecraft [59] or SuperMario [60] were associated with an improvement of hippocampal-associated memory [59].

Some recent and growing papers examined the influence of eSports experience on human development, showing a promising positive relationship between eSports and neural plasticity [61,62]. Moreover, the high degree of participation, the controlled environment, the organized skill ratings, the social nature, and the large amount of data, make the eSport a valid field for neurocognitive research [47]. The metaverse concept can be a promising field in that sense [63].

4. Physical Fitness

Despite the common point of view, eSport is not a passive activity. However, an eAthlete could spend a considerable amount of time in sedentary activities, and video games are a substitute for more intensive exercise because the level of effort achieved is not sufficient to improve cardiorespiratory fitness [64,65].

Indeed, a recent survey observed that almost 90% of elite eAthletes exercise regularly in the gym for about 5 h daily, even more than the physical activity recommendations of WHO for both children and adults [66,67]. Unfortunately, approximately half of the eAthletes investigated did not believe that introducing physical exercise in their training program could have a positive effect on their gaming performance, not recognizing the important relationship between physical and mental health [68]. Conversely, in a more recent survey, Di Francisco et al. reported that 40% of collegiate eAthletes do not participate in any form of physical activity, and Trotter et al. stated that 80% of eSports players do not meet the WHO's physical activity recommendations [4,69].

Therefore, this is a controversial point in the scientific literature, and more studies are needed. "The e-sport study 2019", a study about the demographics and health behavior of eSports players, showed that the professional eAthlete has excellent health and physical status: indeed, eAthletes ranked in the top 10% were more physically active compared to the other players [69].

It is important to screen the physical fitness of these eAthletes. Physical fitness comprises several aspects, including cardiorespiratory fitness (i.e., through cardiopulmonary exercise test with VO2 max evaluation [65,70]), muscle endurance and strength, body composition, balance, and flexibility: testing these is necessary to set a correct training plan, monitor progresses and evaluate physical functioning [71].

The eSports world has enormous potential for physical activity and health-promoting efforts, due to its broad dissemination to a wider public [72]. Indeed, several platforms have developed eGames where level advancement requires the player to walk, run, and ride a bicycle, mixing reality with the virtual world [73]. An interesting recent research shows how eSports could be considered as an exercise prescription in patients with cystic fibrosis [74], to promote physical activity in people with diseases: this is a very important finding as the use of physical activity as a drug against diseases in gaining growing scientific interest [68,75], and therefore eSports can play a role in that field, also though the digital diffusion of physical exercises [76]. Indeed, Tabacof et al. [77] showed how technology and adaptive competitive eSports have the potential to improve social connectedness and inclusion in people with quadriplegia. Moreover, these technologies have been used for other practical tasks such as the mental preparation of athletes for competitions, the creation of rehabilitation exercises for injured athletes or the achievement of some competency goals. Considering the high prevalence of physical inactivity worldwide and the latest diffusion of unhealthy habits also related to the COVID-19 pandemic [41], augmented reality mobile games may provide an enjoyable alternative to engage people who would otherwise never exercise, offering an attractive alternative to traditional physical exercise [78].

5. Cardiovascular System

Despite their growing scientific interest, the physiological demand for eSports has been poorly explored. Cardiovascular adaptations of eSports are surely one of the most debated topics in this context (Figure 1).

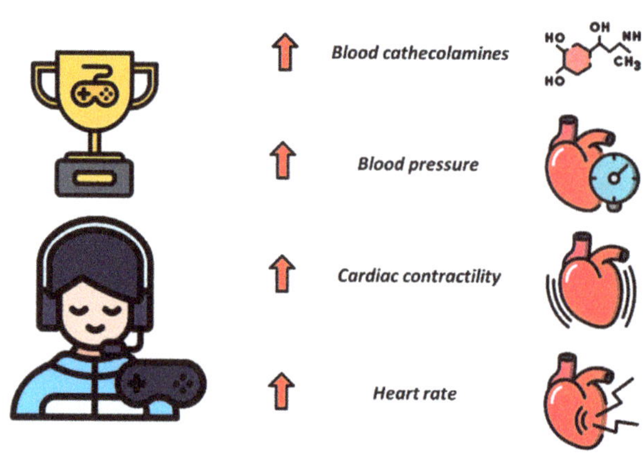

Figure 1. Acute cardiovascular response during eSports session.

An increase in heart rate (HR), blood pressure (BP), and cardiorespiratory fitness during eSports gaming sessions have been observed [79,80]. Like any other skill discipline [81], eSports can cause the release of catecholamines and a consequent rise in BP by augmenting cardiac contractility and increasing HR [60]. The HR increase indicates a parasympathetic withdrawal and/or an upregulation of the sympathetic nervous system possibly through the release of catecholamines. This modulation in eAthletes is particularly evident in some game situations [82]: indeed, HR increases at the beginning of the match or toward the end of a game, and it is higher when the presence of a human opponent is compared with a computer opponent [83]. This means that the sympathetic nervous system in eSports players is activated by competitive play and is modulated by the game situation.

During Fortnite sessions, an eGame in which the player continues playing for as long as possible in an uninterrupted session with increasingly difficult waves of challenges, Valladão et al. [84] described a rise in the mean and peak HR, suggesting that an eAthlete experiences a stressful physiological response.

Moreover, given the rise of systolic BP and HR while playing eSports, other cardiovascular alterations, such as arrhythmias, might be expected to occur in eAthletes. Claire M. Lawley et al. [85] described, in a correspondence published in the New England Journal of Medicine, four instances of syncope in three children due to ventricular tachycardia or ventricular fibrillation while playing electronic games. Mental stress and heightened emotion, such as anger, have been shown to shorten the cardiac ventricular action potential, thus triggering cardiac arrhythmias, in predisposed subjects, such as subjects suffering from long-QT syndrome [86].

Yamagata et al. [87] in their recent article hypothesized three different mechanisms causing the cardiovascular sequelae in eAthletes: sleep deprivation (linked to increased arterial stiffness and endothelial dysfunction, HR variability, cardiac chamber alteration, and cardiac repolarization and increased inflammatory markers), psychological stress (linked to cardiac arrhythmias, increased sympathetic tone) and stimulant use (linked to hypertension, increased level of circulating catecholamines, QT prolongation and coronary artery intimal hyperplasia).

Another potentially worrying cardiovascular aspect is caused by prolonged sitting time. Indeed, water retention can be a consequence of that, measured by bioelectrical impedance and reduced by wearing compression stockings [88]. Deep venous thrombosis is a rare but devastating consequence of that, as gaming has been linked to an hypercoagulability state [89,90]: a prompt diagnosis is crucial when a clinical suspicious raises, with compression ultrasound as a key diagnostic exam [91].

Cardiovascular safety during sports participation aims at avoiding hidden and potentially fatal cardiac events [92,93], through the use of pre-participation cardiovascular strategies, promoted by medical societies [92]. Sudden cardiac death in young athletes is caused by a variety of structural and electrical disorders of the heart, including cardiomyopathies, ion channel disorders, coronary anomalies, and acquired cardiac conditions [92]. Likely, some of the eSports players of all ages have either known or hidden cardiovascular disease. Even minor cardiovascular changes in these people might occasionally have serious consequences [94]. Because it is not feasible to evaluate the entire population that plays video games extensively or intensively, it may be useful to start with professional eAthletes. Therefore, the implementation of a pre-participation evaluation for eAthletes, similar to what is already practiced for skill disciplines athletes, should be considered. A medical evaluation investigating family and personal history, performing a comprehensive physical examination, and using a resting ECG, together with a resting BP value, may be appropriate to identify most cardiovascular conditions that may increase the risk of sudden death in young players. Echocardiography, Holter ECG and magnetic resonance imaging can play a role as a second and third line of athlete's heart diagnosis [95].

6. Conclusions

The media attention and commercial interest in eSports are constantly increasing as well as the number of athletes participating in this discipline. The practice and dedication required for professional eSports are not different from that of any other skill-based discipline. At the same time, an effort must be made to emphasize the importance of practicing more active and intensive physical training accompanied by proper nutrition strategy to maintain eAthletes health. Therefore, there are potential clinical risks for the health of eAthletes associated with eSports practice that has recently become of medical interest. Every gamer who competes or aspires to compete at a high level should be evaluated through a complete pre-participation evaluation (Figure 2) and work on a correct training plan and an injury prevention strategy (Table 2) [96].

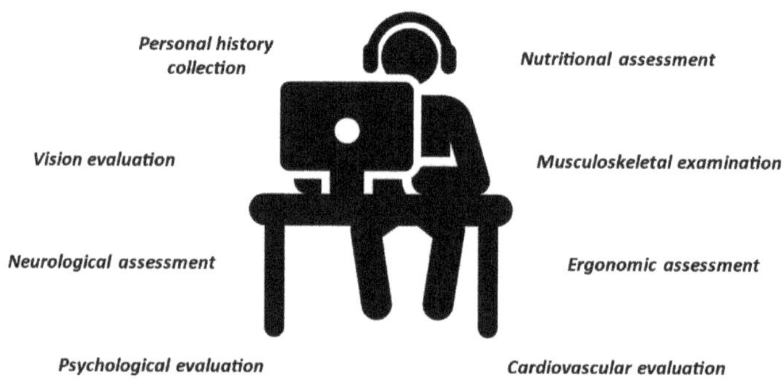

Figure 2. Pre-participation evaluation proposed model for eAthletes.

Table 2. Injury prevention strategies for eAthletes.

Stretching routine before and during gaming
Strengthening exercise
Appropriate posture
Appropriate ergonomics
Limit continuous gaming
Specific eyewear
Maintain volumes in an acceptable range
Adequate level of physical fitness
Adequate level of caloric and water intake
No excessive use of caffeine and energy drinks
Prefer healthy social interactions and team building
Relaxation techniques and meditation
Anxiety and stress control

7. Future Directions

With the help of virtual reality, augmented reality, and artificial intelligence, among other technologies, traditional sports can follow successful strategies from interactive media [79]. Furthermore, specialized monitoring technologies and clinical grade assessments are becoming a topic of growing interest [96]. Future research areas of eSport medicine could be focused on musculoskeletal injuries, ergonomic optimization, vision screening, nutrition strength and conditioning, athlete longevity, performance assessment, and enhancement.

Author Contributions: Conceptualization, S.P. and M.V.; methodology, E.T.; resources, E.C.; data curation, E.B.; writing—original draft preparation, A.R.; writing—review and editing, M.B.; supervision, P.E.A. All authors have read and agreed to the published version of the manuscript.

Funding: This research received no external funding.

Data Availability Statement: Not applicable.

Conflicts of Interest: The authors declare no conflict of interest.

References

1. Hamari, J.; Sjöblom, M. What Is ESports and Why Do People Watch It? *Internet Res.* **2017**, *27*, 211–232. [CrossRef]
2. Rudolf, K.; Bickmann, P.; Froböse, I.; Tholl, C.; Wechsler, K.; Grieben, C. Demographics and Health Behavior of Video Game and ESports Players in Germany: The ESports Study 2019. *Int. J. Environ. Res. Public Health* **2020**, *17*, 1870. [CrossRef] [PubMed]
3. Wattanapisit, A.; Wattanapisit, S.; Wongsiri, S. Public Health Perspectives on ESports. *Public Health Rep.* **2020**, *135*, 295–298. [CrossRef] [PubMed]
4. DiFrancisco-Donoghue, J.; Balentine, J.; Schmidt, G.; Zwibel, H. Managing the Health of the ESport Athlete: An Integrated Health Management Model. *BMJ Open Sport Exerc. Med.* **2019**, *5*, e000467. [CrossRef]
5. Thiel, A.; John, J.M. Is ESport a 'Real' Sport? Reflections on the Spread of Virtual Competitions. *Eur. J. Sport Soc.* **2018**, *15*, 311–315. [CrossRef]
6. Nagorsky, E.; Wiemeyer, J. The Structure of Performance and Training in Esports. *PLoS ONE* **2020**, *15*, e0237584. [CrossRef]
7. Franks, R.R.; King, D.; Bodine, W.; Chisari, E.; Heller, A.; Jamal, F., 4th; Luksch, J.; Quinn, K.; Singh, R.; Solomon, M. AOASM Position Statement on Esports, Active Video Gaming, and the Role of the Sports Medicine Physician. *Clin. J. Sport Med.* **2022**, *32*, e221–e229. [CrossRef]
8. Zwibel, H.; Difrancisco-Donoghue, J.; Defeo, A.; Ii, O.; Yao, S. An Osteopathic Physician's Approach to the Esports Athlete. *J. Am. Osteopath. Assoc.* **2019**, *119*, 756–762. [CrossRef]
9. Anđelić, B.; Bianco, A.; Maksimović, N.; Todorović, N.; Drid, P. A Milestone in the Era of Esports: The Olympics through the Lens of Virtual Reality. *Front. Psychol.* **2022**, *13*, 990189. [CrossRef]
10. Peter, S.C.; Li, Q.; Pfund, R.A.; Whelan, J.P.; Meyers, A.W. Public Stigma Across Addictive Behaviors: Casino Gambling, eSports Gambling, and Internet Gaming. *J. Gambl. Stud.* **2019**, *1*, 247–259. [CrossRef]
11. Kelly, S.; Leung, J. The New Frontier of Esports and Gaming: A Scoping Meta-Review of Health Impacts and Research Agenda. *Front. Sports Act. Living* **2021**, *3*, 640362. [CrossRef] [PubMed]
12. Kane, D.; Spradley, B. Recognizing ESports as a Sport. *Sport J.* **2017**, *19*. Available online: https://thesportjournal.org/article/recognizing-esports-as-a-sport/ (accessed on 20 December 2022).
13. Bányai, F.; Griffiths, M.D.; Király, O.; Demetrovics, Z. The Psychology of Esports: A Systematic Literature Review. *J. Gambl. Stud.* **2019**, *35*, 351–365. [CrossRef] [PubMed]
14. Bányai, F.; Zsila, Á.; Griffiths, M.D.; Demetrovics, Z.; Király, O. Career as a Professional Gamer: Gaming Motives as Predictors of Career Plans to Become a Professional Esport Player. *Front. Psychol.* **2020**, *11*, 1866. [CrossRef] [PubMed]
15. Bonnar, D.; Castine, B.; Kakoschke, N.; Sharp, G. Sleep and Performance in Eathletes: For the Win! *Sleep Health* **2019**, *5*, 647–650. [CrossRef] [PubMed]
16. Dill, K.E.; Dill, J.C. Video Game Violence: A Review of the Empirical Literature. *Aggress. Violent Behav.* **1998**, *3*, 407–428. [CrossRef]
17. Griffiths, M.D.; Kuss, D.J.; King, D.L. Video Game Addiction: Past, Present and Future. *Curr. Psychiatry Rev.* **2012**, *8*, 308–318. [CrossRef]
18. Sirico, F.; Romano, V.; Sacco, A.M.; Belviso, I.; Didonna, V.; Nurzynska, D.; Castaldo, C.; Palermi, S.; Sannino, G.; della Valle, E.; et al. Effect of Video Observation and Motor Imagery on Simple Reaction Time in Cadet Pilots. *J. Funct. Morphol. Kinesiol.* **2020**, *5*, 89. [CrossRef]
19. Lam, W.-K.; Liu, R.-T.; Chen, B.; Huang, X.-Z.; Yi, J.; Wong, D.W.-C. Health Risks and Musculoskeletal Problems of Elite Mobile Esports Players: A Cross-Sectional Descriptive Study. *Sports Med. Open* **2022**, *8*, 65. [CrossRef]
20. Burleigh, T.L.; Griffiths, M.D.; Sumich, A.; Stavropoulos, V.; Kuss, D.J. A Systematic Review of the Co-Occurrence of Gaming Disorder and Other Potentially Addictive Behaviors. *Curr. Addict. Rep.* **2019**, *6*, 383–401. [CrossRef]
21. Louw, S.; Makwela, S.; Manas, L.; Meyer, L.; Terblanche, D.; Brink, Y. Effectiveness of Exercise in Office Workers with Neck Pain: A Systematic Review and Meta-Analysis. *S. Afr. J. Physiother.* **2017**, *73*, 392. [CrossRef] [PubMed]
22. Park, J.-H.M.; Kang, S.-Y.P.; Lee, S.-G.P.; Jeon, H.-S.P. The Effects of Smart Phone Gaming Duration on Muscle Activation and Spinal Posture: Pilot Study. *Physiother. Theory Pract.* **2017**, *33*, 661–669. [CrossRef]
23. Stoffregen, T.A.; Faugloire, E.; Yoshida, K.; Flanagan, M.B.; Merhi, O. Motion Sickness and Postural Sway in Console Video Games. *Hum. Factors* **2008**, *50*, 322–331. [CrossRef] [PubMed]
24. Fiebert, I.; Kistner, F.; Gissendanner, C.; DaSilva, C. Text Neck: An Adverse Postural Phenomenon. *Work* **2021**, *69*, 1261–1270. [CrossRef] [PubMed]

25. Palermi, S.; Massa, B.; Vecchiato, M.; Mazza, F.; de Blasiis, P.; Romano, A.M.; di Salvatore, M.G.; della Valle, E.; Tarantino, D.; Ruosi, C.; et al. Indirect Structural Muscle Injuries of Lower Limb: Rehabilitation and Therapeutic Exercise. *J. Funct. Morphol. Kinesiol.* **2021**, *6*, 75. [CrossRef]
26. Lam, W.-K.; Chen, B.; Liu, R.-T.; Cheung, J.C.-W.; Wong, D.W.-C. Spine Posture, Mobility, and Stability of Top Mobile Esports Athletes: A Case Series. *Biology* **2022**, *11*, 737. [CrossRef]
27. Park, D.-J.; Park, S.-Y. Long-Term Effects of Diagonal Active Stretching versus Static Stretching for Cervical Neuromuscular Dysfunction, Disability and Pain: An 8 Weeks Follow-up Study. *J. Back Musculoskelet. Rehabil.* **2019**, *32*, 403–410. [CrossRef]
28. Geoghegan, L.; Wormald, J.C.R. Sport-Related Hand Injury: A New Perspective of e-Sports. *J. Hand Surg. Eur. Vol.* **2019**, *44*, 219–220. [CrossRef]
29. Wang, D.; Tang, L.; Wu, H.; Gu, D. Analysis of the Effect of Overusing Thumbs on Smartphone Games. *J. Int. Med. Res.* **2019**, *47*, 6244–6253. [CrossRef]
30. Palermi, S.; Annarumma, G.; Spinelli, A.; Massa, B.; Serio, A.; Vecchiato, M.; Demeco, A.; Brugin, E.; Sirico, F.; Giada, F.; et al. Acceptability and Practicality of a Quick Musculoskeletal Examination into Sports Medicine Pre-Participation Evaluation. *Pediatr. Rep.* **2022**, *14*, 207–216. [CrossRef]
31. Belviso, I.; Palermi, S.; Sacco, A.M.; Romano, V.; Corrado, B.; Zappia, M.; Sirico, F. Brachial Plexus Injuries in Sport Medicine: Clinical Evaluation, Diagnostic Approaches, Treatment Options, and Rehabilitative Interventions. *J. Funct. Morphol. Kinesiol.* **2020**, *5*, 22. [CrossRef] [PubMed]
32. Postma, D.B.W.; van Delden, R.W.; van Hilvoorde, I.M. "Dear IOC": Considerations for the Governance, Valuation, and Evaluation of Trends and Developments in ESports. *Front. Sports Act. Living* **2022**, *4*, 899613. [CrossRef] [PubMed]
33. Sheppard, A.L.; Wolffsohn, J.S. Digital Eye Strain: Prevalence, Measurement and Amelioration. *BMJ Open Ophthalmol.* **2018**, *3*, e000146. [CrossRef]
34. Tosini, G.; Ferguson, I.; Tsubota, K. Effects of Blue Light on the Circadian System and Eye Physiology. *Mol. Vis.* **2016**, *22*, 61–72. [PubMed]
35. Emara, A.K.; Ng, M.K.; Cruickshank, J.A.; Kampert, M.W.; Piuzzi, N.S.; Schaffer, J.L.; King, D. Gamer's Health Guide: Optimizing Performance, Recognizing Hazards, and Promoting Wellness in Esports. *Curr. Sports Med. Rep.* **2020**, *19*, 537–545. [CrossRef] [PubMed]
36. Van der Lely, S.; Frey, S.; Garbazza, C.; Wirz-Justice, A.; Jenni, O.G.; Steiner, R.; Wolf, S.; Cajochen, C.; Bromundt, V.; Schmidt, C. Blue Blocker Glasses as a Countermeasure for Alerting Effects of Evening Light-Emitting Diode Screen Exposure in Male Teenagers. *J. Adolesc. Health* **2015**, *56*, 113–119. [CrossRef] [PubMed]
37. Argilés, M.; Quevedo-Junyent, L.; Erickson, G. Topical Review: Optometric Considerations in Sports Versus E-Sports. *Percept. Mot. Ski.* **2022**, *129*, 731–746. [CrossRef]
38. Szot, M.; Frączek, B.; Tyrała, F. Nutrition Patterns of Polish Esports Players. *Nutrients* **2022**, *15*, 149. [CrossRef]
39. Thomas, D.T.; Erdman, K.A.; Burke, L.M. American College of Sports Medicine Joint Position Statement. Nutrition and Athletic Performance. *Med. Sci. Sports Exerc.* **2016**, *48*, 543–568. [CrossRef]
40. Sirico, F.; Fernando, F.; Bianco, A.; Loiacono, C.; Nuccio, F.; Gambardella, F.; Palermi, S.; Montagnani, S.; Biffi, A. Parental Perception of Children's Weight Status: Love Overpasses Scientific Evidence! A Cross-Sectional Observational Study. *High Blood Press. Cardiovasc. Prev.* **2020**, *27*, 29–34. [CrossRef]
41. Palermi, S.; Vecchiato, M.; Pennella, S.; Marasca, A.; Spinelli, A.; de Luca, M.; de Martino, L.; Fernando, F.; Sirico, F.; Biffi, A. The Impact of the COVID-19 Pandemic on Childhood Obesity and Lifestyle-A Report from Italy. *Pediatr. Rep.* **2022**, *14*, 410–418. [CrossRef] [PubMed]
42. DiFrancisco-Donoghue, J.; Werner, W.G.; Douris, P.C.; Zwibel, H. Esports Players, Got Muscle? Competitive Video Game Players' Physical Activity, Body Fat, Bone Mineral Content, and Muscle Mass in Comparison to Matched Controls. *J. Sport Health Sci.* **2022**, *11*, 725–730. [CrossRef] [PubMed]
43. Szot, M.; Karpęcka-Gałka, E.; Dróżdż, R.; Frączek, B. Can Nutrients and Dietary Supplements Potentially Improve Cognitive Performance Also in Esports? *Healthcare* **2022**, *10*, 186. [CrossRef] [PubMed]
44. Mangi, M.A.; Rehman, H.; Rafique, M.; Illovsky, M. Energy Drinks and the Risk of Cardiovascular Disease: A Review of Current Literature. *Cureus* **2017**, *9*, e1322. [CrossRef] [PubMed]
45. Adami, P.E.; Koutlianos, N.; Baggish, A.; Bermon, S.; Cavarretta, E.; Deligiannis, A.; Furlanello, F.; Kouidi, E.; Marques-Vidal, P.; Niebauer, J.; et al. Cardiovascular Effects of Doping Substances, Commonly Prescribed Medications and Ergogenic Aids in Relation to Sports: A Position Statement of the Sport Cardiology and Exercise Nucleus of the European Association of Preventive Cardiology. *Eur. J. Prev. Cardiol.* **2022**, *29*, 559–575. [CrossRef] [PubMed]
46. Humeniuk, R.; Henry-Edwards, S.; Ali, R.; Poznyak, V.; Monteiro, M.G.; World Health Organization. *The Alcohol, Smoking and Substance Involvement Screening Test (ASSIST): Manual for Use in Primary Care*; Addiction Research Center: Baltimore, MD, USA, 2010; p. 68.
47. Campbell, M.J.; Toth, A.J.; Moran, A.P.; Kowal, M.; Exton, C. ESports: A New Window on Neurocognitive Expertise? *Prog. Brain Res.* **2018**, *240*, 161–174. [CrossRef]
48. Palanichamy, T.; Sharma, M.; Sahu, M.; Kanchana, D. Influence of Esports on Stress: A Systematic Review. *Ind. Psychiatry J.* **2020**, *29*, 191–199. [CrossRef]

49. Singh, P.; Sharma, M.K.; Arya, S. Esports and Traditional Sports Players: An Exploration of Psychosocial Profile. *Res. Sq.* **2022**, 1–12. [CrossRef]
50. Della Valle, E.; Palermi, S.; Aloe, I.; Marcantonio, R.; Spera, R.; Montagnani, S.; Sirico, F. Effectiveness of Workplace Yoga Interventions to Reduce Perceived Stress in Employees: A Systematic Review and Meta-Analysis. *J. Funct. Morphol. Kinesiol.* **2020**, *5*, 33. [CrossRef]
51. Triberti, S.; Milani, L.; Villani, D.; Grumi, S.; Peracchia, S.; Curcio, G.; Riva, G. What Matters Is When You Play: Investigating the Relationship between Online Video Games Addiction and Time Spent Playing over Specific Day Phases. *Addict. Behav. Rep.* **2018**, *8*, 185–188. [CrossRef]
52. Buysse, D.J.; Reynolds, C.F., 3rd; Monk, T.H.; Berman, S.R.; Kupfer, D.J. The Pittsburgh Sleep Quality Index: A New Instrument for Psychiatric Practice and Research. *Psychiatry Res.* **1989**, *28*, 193–213. [CrossRef]
53. Weech, S.; Varghese, J.P.; Barnett-Cowan, M. Estimating the Sensorimotor Components of Cybersickness. *J. Neurophysiol.* **2018**, *120*, 2201–2217. [CrossRef]
54. Gao, Z.; Chen, S.; Pasco, D.; Pope, Z. A Meta-Analysis of Active Video Games on Health Outcomes among Children and Adolescents. *Obes. Rev.* **2015**, *16*, 783–794. [CrossRef]
55. Granic, I.; Lobel, A.; Engels, R.C.M.E. The Benefits of Playing Video Games. *Am. Psychol.* **2014**, *69*, 66–78. [CrossRef] [PubMed]
56. Neguț, A.; Jurma, A.M.; David, D. Virtual-Reality-Based Attention Assessment of ADHD: ClinicaVR: Classroom-CPT versus a Traditional Continuous Performance Test. *Child Neuropsychol.* **2017**, *23*, 692–712. [CrossRef] [PubMed]
57. Peñuelas-Calvo, I.; Jiang-Lin, L.K.; Girela-Serrano, B.; Delgado-Gomez, D.; Navarro-Jimenez, R.; Baca-Garcia, E.; Porras-Segovia, A. Video Games for the Assessment and Treatment of Attention-Deficit/Hyperactivity Disorder: A Systematic Review. *Eur. Child Adolesc. Psychiatry* **2022**, *31*, 5–20. [CrossRef]
58. Ashinoff, B.K.; Abu-Akel, A. Hyperfocus: The Forgotten Frontier of Attention. *Psychol. Res.* **2021**, *85*, 1–19. [CrossRef] [PubMed]
59. Stark, C.E.L.; Clemenson, G.D.; Aluru, U.; Hatamian, N.; Stark, S.M. Playing Minecraft Improves Hippocampal-Associated Memory for Details in Middle Aged Adults. *Front. Sports Act. Living* **2021**, *3*, 685286. [CrossRef]
60. Clemenson, G.D.; Stark, C.E.L. Virtual Environmental Enrichment through Video Games Improves Hippocampal-Associated Memory. *J. Neurosci.* **2015**, *35*, 16116–16125. [CrossRef]
61. Monteiro Pereira, A.; Costa, J.A.; Verhagen, E.; Figueiredo, P.; Brito, J. Associations Between Esports Participation and Health: A Scoping Review. *Sports Med.* **2022**, *52*, 2039–2060. [CrossRef]
62. Micallef, D.; Parker, L.; Brennan, L.; Schivinski, B.; Jackson, M. Improving the Health of Emerging Adult Gamers-A Scoping Review of Influences. *Nutrients* **2022**, *14*, 2226. [CrossRef] [PubMed]
63. Cai, L.; Huang, Z.; Feng, Q.; Chang, X.; Yan, K. Co-Transformation of Digital Health and ESport in Metaverse: Moderating Effects of Digital Personality on Mental Health in Multiplayer Online Battle Arena (MOBA). *Int. J. Environ. Res. Public Health* **2022**, *20*, 760. [CrossRef] [PubMed]
64. Segal, K.R.; Dietz, W.H. Physiologic Responses to Playing a Video Game. *Am. J. Dis. Child* **1991**, *145*, 1034–1036. [CrossRef] [PubMed]
65. Biffi, A.; Gallo, G.; Fernando, F.; Sirico, F.; Signorello, M.G.; de Martino, L.; Manole, G.E.; Palermi, S.; Volpe, M. Relationship between Cardiorespiratory Fitness, Baseline Blood Pressure and Hypertensive Response to Exercise in the Ferrari Corporate Population. *High Blood Press. Cardiovasc. Prev.* **2022**, *29*, 81–88. [CrossRef] [PubMed]
66. Kari, T.; Karhulahti, V.M. Do E-Athletes Move? A Study on Training and Physical Exercise in Elite E-Sports. *Int. J. Gaming Comput. Mediat. Simul.* **2017**, *8*, 53–66. [CrossRef]
67. Palermi, S.; Sacco, A.M.; Belviso, I.; Romano, V.; Montesano, P.; Corrado, B.; Sirico, F. Guidelines for Physical Activity—A Cross-Sectional Study to Assess Their Application in the General Population. Have We Achieved Our Goal? *Int. J. Environ. Res. Public Health* **2020**, *17*, 3980. [CrossRef]
68. Lamberti, V.; Palermi, S.; Franceschin, A.; Scapol, G.; Lamberti, V.; Lamberti, C.; Vecchiato, M.; Spera, R.; Sirico, F.; della Valle, E. The Effectiveness of Adapted Personalized Motor Activity (AMPA) to Improve Health in Individuals with Mental Disorders and Physical Comorbidities: A Randomized Controlled Trial. *Sports* **2022**, *10*, 30. [CrossRef]
69. Trotter, M.G.; Coulter, T.J.; Davis, P.A.; Poulus, D.R.; Polman, R. The Association between Esports Participation, Health and Physical Activity Behaviour. *Int. J. Environ. Res. Public Health* **2020**, *17*, 7329. [CrossRef]
70. Ortolan, S.; Neunhaeuserer, D.; Battista, F.; Patti, A.; Gobbo, S.; di Bella, C.; Quinto, G.; Vecchiato, M.; Frigo, A.C.; Bergamin, M.; et al. Physical Fitness as a Prognostic Marker for Infectious Events in Kidney Transplant Recipients. *Am. J. Nephrol.* **2022**, *53*, 1–9. [CrossRef]
71. De Oliveira Tavares, V.D.; Vancampfort, D.; Hallgren, M.; Heissel, A.; Chaparro, C.G.A.P.; Solmi, M.; Tempest, G.D.; de Oliveira Neto, L.; Galvão-Coelho, N.L.; Firth, J.; et al. Reliability and Validity of Physical Fitness Tests in People with Mental Disorders: A Systematic Review and Meta-Analysis. *Physiother. Res. Int.* **2021**, *26*, e1904. [CrossRef]
72. Ketelhut, S.; Martin-Niedecken, A.L.; Zimmermann, P.; Nigg, C.R. Physical Activity and Health Promotion in Esports and Gaming-Discussing Unique Opportunities for an Unprecedented Cultural Phenomenon. *Front. Sports Act. Living* **2021**, *3*, 693700. [CrossRef] [PubMed]
73. Xian, Y.; Xu, H.; Xu, H.; Liang, L.; Hernandez, A.F.; Wang, T.Y.; Peterson, E.D. An Initial Evaluation of the Impact of Pokémon GO on Physical Activity. *J. Am. Heart Assoc.* **2017**, *6*, e005341. [CrossRef] [PubMed]

74. Campos, N.E.; Heinzmann-Filho, J.P.; Becker, N.A.; Schiwe, D.; Gheller, M.F.; de Almeida, I.S.; Donadio, M.V.F. Evaluation of the Exercise Intensity Generated by Active Video Gaming in Patients with Cystic Fibrosis and Healthy Individuals. *J. Cyst. Fibros.* **2020**, *19*, 434–441. [CrossRef]
75. Compagno, S.; Palermi, S.; Pescatore, V.; Brugin, E.; Sarto, M.; Marin, R.; Calzavara, V.; Nizzetto, M.; Scevola, M.; Aloi, A.; et al. Physical and Psychological Reconditioning in Long COVID Syndrome: Results of an out-of-Hospital Exercise and Psychological—Based Rehabilitation Program. *IJC Heart Vasc.* **2022**, *41*, 101080. [CrossRef]
76. Solis-Navarro, L.; Gismero, A.; Fernández-Jané, C.; Torres-Castro, R.; Solá-Madurell, M.; Bergé, C.; Pérez, L.M.; Ars, J.; Martín-Borràs, C.; Vilaró, J.; et al. Effectiveness of Home-Based Exercise Delivered by Digital Health in Older Adults: A Systematic Review and Meta-Analysis. *Age Ageing* **2022**, *51*, afac243. [CrossRef] [PubMed]
77. Tabacof, L.; Dewil, S.; Herrera, J.E.; Cortes, M.; Putrino, D. Adaptive Esports for People with Spinal Cord Injury: New Frontiers for Inclusion in Mainstream Sports Performance. *Front. Psychol.* **2021**, *12*, 612350. [CrossRef]
78. Bock, B.C.; Dunsiger, S.I.; Ciccolo, J.T.; Serber, E.R.; Wu, W.-C.; Tilkemeier, P.; Walaska, K.A.; Marcus, B.H. Exercise Videogames, Physical Activity, and Health: Wii Heart Fitness: A Randomized Clinical Trial. *Am. J. Prev. Med.* **2019**, *56*, 501–511. [CrossRef]
79. Goldfield, G.S.; Kenny, G.P.; Hadjiyannakis, S.; Phillips, P.; Alberga, A.S.; Saunders, T.J.; Tremblay, M.S.; Malcolm, J.; Prud'homme, D.; Gougeon, R.; et al. Video Game Playing Is Independently Associated with Blood Pressure and Lipids in Overweight and Obese Adolescents. *PLoS ONE* **2011**, *6*, e26643. [CrossRef]
80. Siervo, M.; Sabatini, S.; Fewtrell, M.S.; Wells, J.C.K. Acute Effects of Violent Video-Game Playing on Blood Pressure and Appetite Perception in Normal-Weight Young Men: A Randomized Controlled Trial. *Eur. J. Clin. Nutr.* **2013**, *67*, 1322–1324. [CrossRef]
81. Pelliccia, A.; Sharma, S.; Gati, S.; Bäck, M.; Börjesson, M.; Caselli, S.; Collet, J.-P.; Corrado, D.; Drezner, J.A.; Halle, M.; et al. 2020 ESC Guidelines on Sports Cardiology and Exercise in Patients with Cardiovascular Disease. *Eur. Heart J.* **2021**, *42*, 17–96. [CrossRef]
82. Sousa, A.; Ahmad, S.L.; Hassan, T.; Yuen, K.; Douris, P.; Zwibel, H.; DiFrancisco-Donoghue, J. Physiological and Cognitive Functions Following a Discrete Session of Competitive Esports Gaming. *Front. Psychol.* **2020**, *11*, 1030. [CrossRef] [PubMed]
83. Watanabe, K.; Saijo, N.; Minami, S.; Kashino, M. The Effects of Competitive and Interactive Play on Physiological State in Professional Esports Players. *Heliyon* **2021**, *7*, e06844. [CrossRef] [PubMed]
84. ValladÃo, S.P.; Middleton, J.; Andre, T.L. Esport: Fortnite Acutely Increases Heart Rate of Young Men. *Int. J. Exerc. Sci.* **2020**, *13*, 1217–1227. [PubMed]
85. Lawley, C.M.; Skinner, J.R.; Turner, C. Syncope Due to Ventricular Arrhythmia Triggered by Electronic Gaming. *N. Engl. J. Med.* **2019**, *381*, 1180–1181. [CrossRef]
86. Priori, S.G.; Wilde, A.A.; Horie, M.; Cho, Y.; Behr, E.R.; Berul, C.; Blom, N.; Brugada, J.; Chiang, C.-E.; Huikuri, H.; et al. HRS/EHRA/APHRS Expert Consensus Statement on the Diagnosis and Management of Patients with Inherited Primary Arrhythmia Syndromes: Document Endorsed by HRS, EHRA and APHRS in May 2013 and by ACCF, AHA, PACES and AEPC in June 2013. *Heart Rhythm.* **2013**, *10*, 1932–1963. [CrossRef]
87. Yamagata, K.; Yamagata, L.M.; Abela, M. A Review Article of the Cardiovascular Sequalae in Esport Athletes: A Cause for Concern? *Hell. J. Cardiol.* **2022**, *68*, 40–45. [CrossRef]
88. Schmidt, S.C.E.; Sell, S.; Woll, A. The Use of Compression Stockings to Reduce Water Retention in the Legs During Gaming and Esports: Randomized Controlled Field Study. *Interact. J. Med. Res.* **2022**, *11*, e25886. [CrossRef]
89. Rambaran, K.A.; Alzghari, S.K. Gamer's Thrombosis: A Review of Published Reports. *Ochsner J.* **2020**, *20*, 182–186. [CrossRef]
90. Chang, H.-C.L.; Burbridge, H.; Wong, C. Extensive Deep Vein Thrombosis Following Prolonged Gaming ('Gamer's Thrombosis'): A Case Report. *J. Med. Case Rep.* **2013**, *7*, 235. [CrossRef]
91. Di Vilio, A.; Vergara, A.; Desiderio, A.; Iodice, F.; Serio, A.; Palermi, S.; Gambardella, F.; Sperlongano, S.; Gioia, R.; Acitorio, M.; et al. Incremental Value of Compression Ultrasound Sonography in the Emergency Department. *World J. Crit. Care Med.* **2021**, *10*, 194–203. [CrossRef]
92. Vessella, T.; Zorzi, A.; Merlo, L.; Pegoraro, C.; Giorgiano, F.; Trevisanato, M.; Viel, M.; Formentini, P.; Corrado, D.; Sarto, P. The Italian Preparticipation Evaluation Programme: Diagnostic Yield, Rate of Disqualification and Cost Analysis. *Br. J. Sports Med.* **2020**, *54*, 231–237. [CrossRef] [PubMed]
93. Palermi, S.; Sirico, F.; Fernando, F.; Gregori, G.; Belviso, I.; Ricci, F.; D'Ascenzi, F.; Cavarretta, E.; de Luca, M.; Negro, F.; et al. Limited Diagnostic Value of Questionnaire-Based Pre-Participation Screening Algorithms: A "Risk-Exposed" Approach to Sports Activity. *J. Basic Clin. Physiol. Pharm.* **2022**, *33*, 655–663. [CrossRef] [PubMed]
94. Gwinup, G.; Haw, T.; Elias, A. Cardiovascular Changes in Video-Game Players. *Postgrad. Med.* **1983**, *74*, 245–248. [CrossRef]
95. D'Andrea, A.; Sperlongano, S.; Russo, V.; D'Ascenzi, F.; Benfari, G.; Renon, F.; Palermi, S.; Ilardi, F.; Giallauria, F.; Limongelli, G.; et al. The Role of Multimodality Imaging in Athlete's Heart Diagnosis: Current Status and Future Directions. *J. Clin. Med.* **2021**, *10*, 5126. [CrossRef] [PubMed]
96. Silva, A.S.; Correia, M.V.; Silva, H.P. Invisible ECG for High Throughput Screening in ESports. *Sensors* **2021**, *21*, 7601. [CrossRef]

sports

Article

Identification of Optimal Movement Patterns for Energy Pumping

Micha Luginbühl [1], Micah Gross [2], Silvio Lorenzetti [2,3], David Graf [4] and Martin J. Bünner [1,*]

1 Institute of Computational Engineering, University of Applied Sciences OST, Werdenbergstr. 4, CH-9470 Buchs, Switzerland; micha.luginbuehl@ost.ch
2 Swiss Federal Institute of Sport (SFISM), CH-2532 Magglingen, Switzerland; micah.gross@baspo.admin.ch (M.G.); silvio.lorenzetti@baspo.admin.ch (S.L.)
3 Institute of Biomechanics, D-HEST, ETHZ, CH-8092 Zurich, Switzerland
4 Swiss Cycling, Sportstrasse 44, CH-5240 Grenchen, Switzerland; david.graf@swiss-cycling.ch
* Correspondence: martin.buenner@ost.ch

Abstract: Energy pumping is a way to gain kinetic energy based on an active vertical center of mass movement in rollers in sports like skateboarding, skicross, snowboard cross and BMX. While the principle of the energy transfer from the vertical movement to the horizontal movement is well understood, the question of how to achieve the optimal energy transfer is still unresolved. In this paper, we introduce an inverse pendulum model to describe the movement of the center of mass of an athlete performing energy pumping. On this basis, the problem of identifying the optimal movement pattern is formulated as an optimal control problem. We solve the discretized optimal control problem with the help of a SQP-algorithm. We uncover that the optimal movement pattern consists of a jumping, flying, and landing phase, which has to be timed precisely. We investigate how the maximal horizontal speed depends on parameters like rollers height and maximal normal force of the athlete. Additionally, we present a qualitative comparison of our results with measured results from BMX-racing. For athletes and coaches, we advice on the basis of our results that athlete's performance is optimized by using maximal force and adopt an exact and proper timing of the movement pattern.

Keywords: energy pumping; ski-cross; BMX-racing; optimal control theory; SQP-algorithm; center of mass movement; rollers; digital technology; sports

check for updates

Citation: Luginbühl, M.; Gross, M.; Lorenzetti, S.; Graf, D.; Bünner, M.J. Identification of Optimal Movement Patterns for Energy Pumping. *Sports* **2023**, *1*, 0. https://doi.org/

Academic Editors: Flávio Antônio de Souza Castro, Rui Miguel Simões de Azevedo and Rodrigo Zacca

Received: 31 October 2022
Revised: 17 January 2023
Accepted: 23 January 2023
Published: 30 January 2023

1. Introduction

In many sports like skateboarding, BMX, ski/snowboard cross, the racing track includes wave wells, moguls or rollers. The aim for the athlete is to loose as little time as possible and exit the rollers with a high speed. Here, the potential energy as well as specific movements of the center of mass (energy pumping) can be used to gain speed. For BMX racing the surprisingly high importance of energy pumping has been demonstrated in experiments [1].

Here we define energy pumping as all active and targeted movements of the center of mass of the athlete to generate kinetic energy in sports like skateboarding, skicross, snowboard cross, BMX and gymnastics. The energy transfer is common in sports eq. gymnastics [2] or sailing [3]. Also a swing can be pumped with vertical movements of the center of mass [4,5]. While descending a slope, the force in the direction of the slope can be used for acceleration and results in a larger speed. This accelerating force can be enlarged by a rise of the normal force during vertical acceleration of the center of mass.

In recent years, some progess has been made by formulating biomechanical problems as optimal control problems [6–9]. In this paper, the problem of identifying the optimal movement pattern for athlete's energy pumping on a one-dimensional track is formulated as an optimal control problem, which is standard in the framework of optimal control theory [10]. The discretized optimal control problem turns out to be a high-dimensional nonlinear program (NLP). In this formulation, the athlete's optimal movement pattern is a solution of the KKT-equations [11], derived from the NLP, in a high-dimensional real-valued space. The task to find these KKT-points in high-dimensional spaces is in practice extremely challenging. More precisely, it is far beyond what humans can do without help.

Since not so long ago, the SQP-algorithm, as proposed by Han, and Powell [12,13], and Schittkowski [14], has been shown to be able to identify KKT-points in the high-dimensional real-valued search space. In practice, the SQP-algorithm, running on a computer, can find KKT-points in spaces with several hundred or thousands dimensions. But, the effort to find these solutions is tremendous and without the help of computers, we were not able to find those. Also the optimal movement patterns, which will be presented in this paper, would not be accessible without the help of computers, which perform step-by-step intelligent numerical mathematics procedures. As such, the SQP-method is considered for good reasons as one tool out of the rich toolbox of artificial intelligence.

Historically, as the SQP-method was increasingly available to the scientific and engineering community and the 1990's and 2000's, it enlarged the set of problems, which can be solved. No wonder that the SQP-method was used in several engineering disciplines, like radio frequency design [15], wind turbine airfoil design [16], turbine design [17], mechanical engineering [18], and 2D airfoil design [19], to identify optimal designs, not known to engineers until then.

The optimal vertical movement during energy pumping including friction and different forms of moguls is not well understood. Therefore the aim of this work is to build a model based on the law of motions and find the optimal path.

2. The Inverted Pendulum Model and the Optimal Control Problem

2.1. The Inverted Pendulum Model

Inverted-pendulum models are proven tools as a simplified description of complex human movements [20,21]. To our knowledge, an inverted-pendulum-model for the energy pumping performed by an athlete has not been reported elsewhere. Therefore, we will present the model in detail in this section. We consider a one-dimensional track of length L with a height profile $z = u(x), 0 \leq x \leq L$. The athlete moves in positive x-direction. The local slope of the track is $u'(x)$ and the angle between the x-axis and the track is $\phi(x) = \arctan(u'(x))$. The vector $\vec{N}$ is the upwards-pointing normal vector on the track $u(x)$:

$$\vec{N}(x) = (\, -\sin(\phi(x)), \quad \cos(\phi(x))\,), \tag{1}$$

which we will call *track normal unit vector* in the following.

We model the athlete, including his equipment, as an inverted pendulum, moving from $x = 0$ to $x = L$:

1. The athlete, including his or her equipment, has a single contact point

   $$\vec{s}(t) = (\, x(t), \quad u(x(t))\,), \tag{2}$$

   at which he is in contact with the track for all times (see Figure 1). From now on, we will write athlete only, instead of athlete, including his or her equipment for simplicity. The athlete's mass is concentrated in the center of mass (COM) located at

   $$\vec{r}(t) = (\, r_x(t), \quad r_z(t)\,). \tag{3}$$

2. The body axis of the athlete is modelled as a straight line between the contact point positioned at $\vec{s}(t)$ and the COM positioned at $\vec{r}(t)$. The straight line is considered as a stiff bar in the sense that forces can be transferred to the COM from the contact point along the straight line.
3. The body axis is always perpendicular to the track. Therefore, the vector $\vec{h}(t) = \vec{s}(t) - \vec{r}(t)$ is perpendicular to the track, and parallel to $\vec{N}(x(t))$. The length of $\vec{h}(t)$

we call effective body height: $h(t) = |\vec{h}(t)|$. With this, the location of the COM can be computed from the location of the contact point and the effective body height: $\vec{r}(t) = \vec{s}(t) + h(t) \cdot N(x(t))$. As the athlete bends and stretches his body, he or she can vary his or her effective body height between the minimum effective body height h_{min} (body bent maximally) to the maximal effective body height h_{max} (body stretched maximally).

4. As the athlete moves along the track, a normal force acts from the track to the athlete's COM. This force acts along the body axis, from the contact point to the COM. Therefore, the athlete in our model has to make sure that his or her body axis is exactly parallel to the normal forces. This is in contrast to some sports like cycling, or skiing, there the athlete can balance forces by leaning backward or forward, since it equipment (ski, bike) provides an extended contact line, or several isolated contact points, between the athlete and the track.

With this, we arrive at a 2-dimensional dynamical problem for the movement of the COM with:

$$\vec{v} = \dot{\vec{r}}, \quad \vec{a} = \ddot{\vec{r}}, \quad v = |\vec{v}|, \quad a = |\vec{a}|. \tag{4}$$

Additionally, we introduce (a) the trajectory tangential unit vector $\vec{T}(\dot{\vec{r}})$, which is tangential to the movement of the COM for all times, and (b) the trajectory normal unit vector $\vec{R}(\dot{\vec{r}}, \ddot{\vec{r}})$, which points from the COM to the middle of the curvature circle.

$$\vec{T}(\dot{\vec{r}}) = \frac{\vec{v}}{v}, \tag{5}$$

$$\vec{R}(\dot{\vec{r}}, \ddot{\vec{r}}) = \frac{\dot{\vec{T}}}{|\dot{\vec{T}}|}, \quad |\dot{\vec{T}}| \neq 0. \tag{6}$$

The curvature of the COM's path, defined as the inverse of the curvature radius, is:

$$k(\dot{\vec{r}}, \ddot{\vec{r}}) = \frac{|\vec{v} \times \vec{a}|}{v^3} \tag{7}$$

Now, let us consider the forces, respectively accelerations, acting on the athlete's COM. These are gravity, friction, body force, and centrifugal forces. The gravity force, respectively acceleration is (see Figure 2):

$$\vec{F}_G = \begin{pmatrix} 0 \\ -mg \end{pmatrix}, \quad \vec{a}_G = \begin{pmatrix} 0 \\ -g \end{pmatrix} \tag{8}$$

The centrifugal force, respectively acceleration is given by:

$$\vec{F}_Z(\dot{\vec{r}}, \ddot{\vec{r}}) = mv^2 k(\dot{\vec{r}}, \ddot{\vec{r}}) \vec{R}(\dot{\vec{r}}, \ddot{\vec{r}}), \qquad \vec{a}_Z(\dot{\vec{r}}, \ddot{\vec{r}}) = v^2 k(\dot{\vec{r}}, \ddot{\vec{r}}) \vec{R}(\dot{\vec{r}}, \ddot{\vec{r}}). \tag{9}$$

By stretching or folding his or her body along the body axis, the athlete is able to apply an additional force to the movement of the COM, which we call body force from now on. In the model presented, the athlete can only provide forces parallel to his or her body axis. Hence, the body force acts in the direction of the track normal unit vector $\vec{N}$.

$$\vec{F}_N(t, \vec{r}) = F_N(t) \cdot \vec{N}(\vec{r}) = m \cdot a_N(t) \cdot \vec{N}(\vec{r}). \tag{10}$$

For any time, we assume: (a) the body acceleration $a_N(t)$ cannot be negative, and (b) the body acceleration cannot be larger than a maximal value a_N^{max}, since the physical strength of the athlete is finite. Therefore, the restriction: $0 \leq a_N(t) \leq a_N^{max}$ holds for all times.

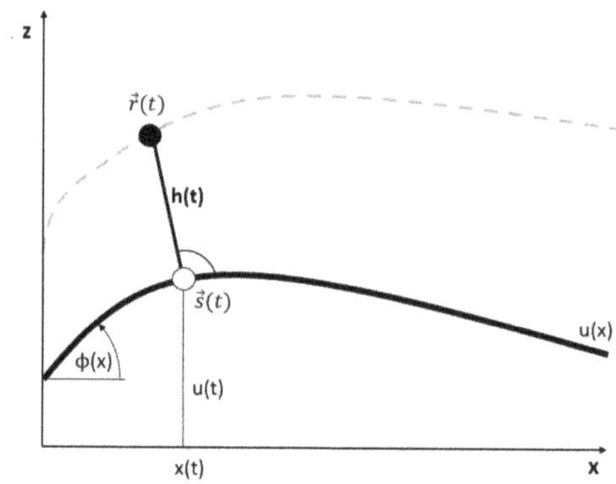

Figure 1. Inverted pendulum model for athlete's movement along the track with height profile $u(x)$ (thick line). Dashed line: Trajectory of the COM (full circle). The contact point is shown as an open circle.

In the model we introduce two friction forces, the air friction and the sliding or rolling friction:

$$\vec{F}_R(t,\vec{r},\dot{\vec{r}}) = -[\mu F_N(t,\vec{r}) + m\alpha(\vec{r},\dot{\vec{r}})]\vec{T}(\dot{\vec{r}}). \tag{11}$$

The air friction is modelled with a simple Newtonian ansatz leading to frictional force with a quadratic law in v:

$$\alpha(\vec{r},\dot{\vec{r}}) = \frac{c_w \rho A(\vec{r}) v^2}{2m}, \tag{12}$$

with c_w athlete's c_w-value, ρ density of the medium (in our case: air), and a time-dependant cross-sectional area $A(t)$ of the athlete. We model the variation of the cross-sectional area with the effective body height with the help of a simple linear law:

$$A(\vec{r}) = \frac{A_{max}}{h_{max}} \cdot h(\vec{r}), \tag{13}$$

with A_{max} being the athlete's cross-sectional with a fully stretched body. The sliding or rolling friction is modelled to be proportional to the normal force with the coefficient μ.

We are now in the position to establish the equations of motion for the movement of the COM with the help of Newton's law:

$$m\ddot{\vec{r}} = \vec{F}_R(t,\vec{r},\dot{\vec{r}}) + \vec{F}_G + \vec{F}_N(t,\vec{r}) + \vec{F}_Z(\dot{\vec{r}},\ddot{\vec{r}}), \tag{14}$$

As a result we can derive an implicit non-autonomous set of second order differential equations for the movement of the COM, by applying the expressions for the accelerations as derived earlier in this paper:

$$\ddot{\vec{r}} = -[\mu a_N(t) + \alpha(\vec{r},\dot{\vec{r}})]\vec{T}(\dot{\vec{r}}) + v^2 k(\dot{\vec{r}},\ddot{\vec{r}})\vec{R}(\dot{\vec{r}},\ddot{\vec{r}}) + a_N(t) \cdot \vec{N}(\vec{r}) + \vec{a}_G, \tag{15}$$

with either:

1. initial conditions:

$$\dot{\vec{r}}(0) = \vec{v}_0, \qquad \vec{r}(0) = \vec{r}_0, \tag{16}$$

2. or circular conditions:

$$\dot{\vec{r}}(0) = \dot{\vec{r}}(\tau), \qquad \vec{r}(0) = \vec{r}(\tau). \tag{17}$$

With circular conditions (17), we model a situation in which the athlete moves on a periodic track, where the shape of the track repeats itself over-and-over again (with infinite length). With initial conditions (16), we model a situation in which the athlete moves on a track with finite length L only once.

Note the implicitness of the set of differential Equation (15), since $\ddot{\vec{r}}$ appears on the left-hand side and the right-hand side of the equations. Therefore, straightforward ODE-solvers like Runge-Kutta-schemes, implementing a discretization in time, cannot be used for the numerical solution. As we will solve (15) as constraints to an optimization problem (to be shown later in this paper), the implicit nature of (15) is taken care of in a straightforward way.

We are interested in the solution of the set of ODEs for $0 \leq t \leq \tau$ with $x(\tau) = L$. For initial conditions, τ is the time to complete the track of length L. For circular conditions, τ is the time to complete a single period of length L of the track (which has infinite length).

Note that the body acceleration $a_N(t)$ acts as an external driver, or external force, on the ODE-system (15) and needs to be fixed prior to the solution. As such COM's trajectory $\vec{r} = \vec{r}[a_N]$ is a functional of $a_N(t)$. The same is true for the completion time $\tau = \tau[a_N]$.

It is important to note, that not all set of parameters, initial conditions, and body accelerations $a_N(t)$ lead to solutions, which satisfiy the implicit constraint $h_{min} \leq h(t) \leq h_{max}$. Therefore, we restrict the set of parameters, initial conditions, and body accelerations $a_N(t)$ to those who give rise to solutions which satisfy the implicit constraint. $h_{min} \leq h(t) \leq h_{max}$.

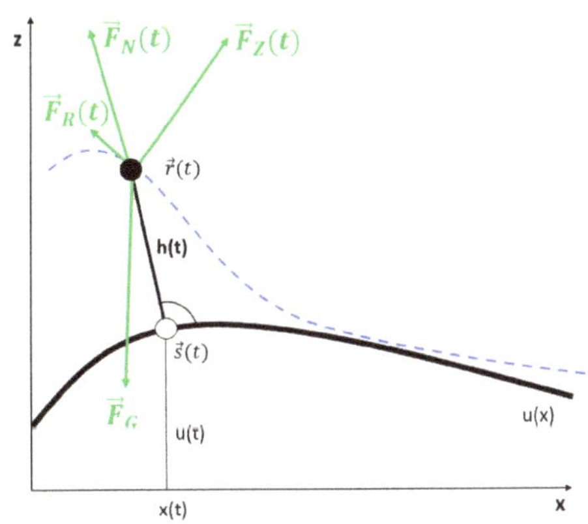

Figure 2. The movement of the COM (full circle) along his continuous path (dashed line) of the movement of the COM (full circle). The contact point is shown as an open circle.

2.2. The Optimal Control Problem

The aim of this paper is to identify the optimal movement pattern for energy pumping under variation of the relevant parameters such aus athlete's mass, friction or the shape of the track. The optimal movement pattern is defined as the one, which results in the fastest completion of the track, or, which gives the smallest overall time for track completion τ.

The athlete's degree of freedom to modify the movement pattern is the acceleration $a_N(t)$, which encompasses the body movement of the athlete, including the variation of athlete's cross sectional area $A(t) = A[a_N](t)$, since the latter is also a functional of $a_N(t)$. This leads to a time-continuous optimal control problem with the function $a_N(t)$ as functional design variable:

$$
\begin{aligned}
\tau^* &= \min_{a_N(t)} \tau[a_N], && (18)\\
h_{min} &\le h[a_N](t) \le h_{max}, \qquad 0 \le t \le \tau, && (19)\\
0 &\le a_N(t) \le a_N^{max}, \qquad 0 \le t \le \tau, && (20)\\
\ddot{\vec{r}} &= -[\mu a_N(t) + \alpha(\vec{r},\dot{\vec{r}})]\vec{T}(\dot{\vec{r}}) + v^2 k(\dot{\vec{r}},\ddot{\vec{r}})\vec{R}(\dot{\vec{r}},\ddot{\vec{r}}) + a_N(t)\cdot\vec{N}(\vec{r}) + \vec{a}_G, && (21)\\
x(\tau) &= L && (22)
\end{aligned}
$$

Additionally, we apply either initial value conditions (16) or circular conditions (17) to the ODE.

In less mathematical terms, the meaning of the optimization problem (18)–(22) is: Find the time-dependant body acceleration $a_N(t)$ such that the track $u(x)$ is completed as fast as possible starting from $x = 0$ to $x = L$, taking into account all restrictions imposed on the problem. The solution of the optimization problem delivers the optimal acceleration $a_N^*(t)$, the minimal possible completion time τ^*, as well as the optimal trajectory of the COM of the athlete $(\dot{\vec{r}}^*, \vec{r}^*)$. All three together, forces and trajectory, defines the optimal movement pattern of the athlete.

In order to solve the optimal control problem, we discretize the model in space. To this end, we introduce the equidistant discretization of the x-coordinate of the contact point with an interval Δx: $x_i^s = (i-1)\Delta x, i = 1, \dots N$ with $N = \frac{L}{\Delta x} + 1$. The track is discretized as $\vec{s}_i = (x_i^s, \quad u(x_i^s))$, also $\phi_i = \arctan(u'(x_i^s))$ and the discretized normal vector is $\vec{N}_i = (-\sin(\phi_i), \quad \cos(\phi_i))$. Accordingly, discretizations apply as: $h_i = h(x_i^s)$ und $\vec{r}_i = \vec{s}_i + h_i \cdot \vec{N}_i$. As such, the continuous movement of the COM $\vec{r}(t)$ is approximated by an polygonal line (see Figure 3).

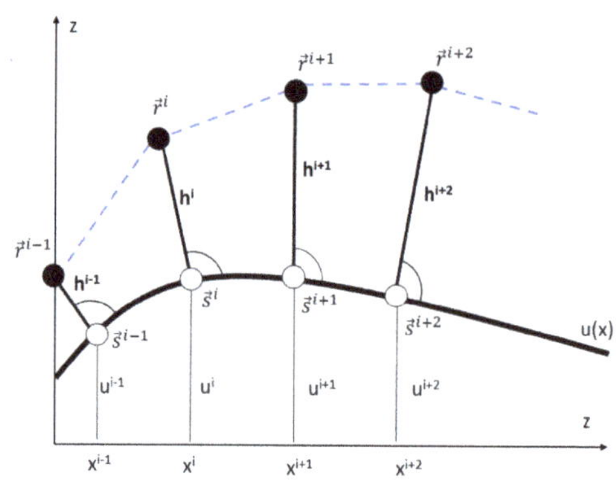

Figure 3. Polygonal line (dashed line) of the movement of the COM (full circle). The base points are shown as open circles.

On this basis, we define $(3N-1)$ design variables (h^i, v^i, a_N^i) for the time-discrete optimal control problem, with $h^i = (h^1, \dots, h^N), v^i = (v^1, \dots, v^{N-1}), a_N^i = (a_N^1, \dots, a_N^N)$. Here, v^i is the magnitude of the velocity as the COM moves from $\vec{r}^i$ to $\vec{r}^{i+1}$. From the design variables together with the parameters all dynamical variables can be computed. The discretized version of accelerations (a_x^i, a_z^i) are computed with the help of first-order finite differences from the velocities. With this, the discretized versions of the tangential and radial vectors are computed. The time for the COM to move from i to $(i+1)$ is:

$$
\Delta\tau^i = \frac{|\vec{r}^{i+1} - \vec{r}^i|}{v^i} \tag{23}
$$

The overall time τ with $x(\tau) = L$, which is needed to complete the track, can be computed as a sum of $N-1$ time steps:

$$\tau(h^i, v^i) = \sum \Delta\tau^i = \sum \frac{|\vec{r}^{i+1} - \vec{r}^i|}{v^i} \quad (24)$$

As a result, the optimization problem reads:

$$\begin{aligned}
\tau^* &= \min_{(h^i, v^i, a_N^i)} \tau(h^i, v^i), && (25)\\
h_{min} &\leq h^i \leq h_{max}, \quad i = 1, \dots N && (26)\\
a_N^{min} &\leq a_N^i \leq a_N^{max}, \quad i = 1, \dots N && (27)\\
v^i &\geq 0, \quad i = 1, \dots N-1 && (28)\\
a_x^i &= -[\mu a_N^i + \alpha(\vec{r}^i, \dot{\vec{r}}^i)]\vec{T}_x(\dot{\vec{r}}^i) + (v^i)^2 k(\dot{\vec{r}}^i, \ddot{\vec{r}}^i)\vec{R}_x(\dot{\vec{r}}^i, \ddot{\vec{r}}^i) + a_N^i \cdot \vec{N}_x(\vec{r}^i), && (29)\\
a_z^i &= -[\mu a_N^i + \alpha(\vec{r}^i, \dot{\vec{r}}^i)]\vec{T}_z(\dot{\vec{r}}^i) + (v^i)^2 k(\dot{\vec{r}}^i, \ddot{\vec{r}}^i)\vec{R}_z(\dot{\vec{r}}^i, \ddot{\vec{r}}^i) + a_N^i \cdot \vec{N}_z(\vec{r}^i) + g, && (30)
\end{aligned}$$

In case of simple initial conditions, we additionally have the constraints:

$$\begin{aligned}
v^1 &= v_0, && (31)\\
h^1 &= h_0. && (32)
\end{aligned}$$

In case of circular initial conditions, we additionally have the constraints:

$$\begin{aligned}
v^1 &= v^{N-1}, && (33)\\
h^1 &= h^{N-1}, && (34)\\
h^2 &= h^N. && (35)
\end{aligned}$$

In case of simple initial conditions, the optimization problem is an NLP with $3N - 1$ real-valued design variables, $5N$ inequality constraints and $2N + 2$ equality constraints. In case of simple initial conditions, the optimization problem is an NLP with real-valued $3N - 1$ design variables, $6N$ inequality constraints and $2N + 3$ equality constraints.

3. Results

3.1. Solution of the Discrete Optimal Control Problem with a SQP-Algorithm

At first, we present the solution of the discretized optimal control problem (25)–(30) with the help of a SQP-algorithm as proposed by Han, and Powell [12,13]. We use circular conditions (17), with the length of a single period of the track $L = 6.0$ m. The height profile is a simple cosinus-curve with a height difference of 0.2 m between highest and lowest point: $u(x) = (1 - \cos(\frac{\pi x}{3})) \cdot 0.1$ m, and $\rho = 1.225 \frac{\text{kg}}{\text{m}^3}$ For the athlete's parameters we take: $m = 75$ kg; $A^{max} = 1.1$ m^2; $c_w = 0.3$; $\mu = 0.02$; $a_N^{min} = 0$; $a_N^{max} = 2g$; $h^{min} = 0.7$ m; $h^{max} = 1.2$ m with $g = 9.81 \frac{\text{m}}{\text{s}^2}$.

We choose an interval of $\Delta x = 0.1$ m for the spatial discretization step, and therefore, $N = 61$. On this basis, the NLP (25)–(30) is formulated with 182 design variables, 305 linear inequality constraints, and 124 nonlinear equality constraints. It is important for us to highlight the following: (a) with 182 design variables the search is performed in a 182-real-valued space, which is a tremendously large search space. (b) The 124 nonlinear equality constraints restrict the feasible region to a 62-dimensional subspace. Even if this sounds large, in fact, the feasible region is much, much smaller compared to the search space. (c) As a result, we expect it to be non-trivial to find the small feasible solution of the optimal control problem starting from an infeasible initial value.

Not surprisingly, it turned out, that random initial values for the search in the 182-dimensional space are in almost any case too far away from the feasible region, such that the SQP-algorithm was unable to find the feasible region, and therefore did not manage to solve the problem. Alternatively, we used random initial values of the form $u_0 = (h_0 \cdot (1, 1, \dots 1), v_0 \cdot (1, 1, \dots 1), a_{N,0} \cdot (1, 1, \dots 1))$ with random numbers h_0, v_0, and $a_{N,0}$. With this, we solved the NLP with a set 1000 random initial conditions. From those 869 SQP-runs did not converge as the algorithm was unable to find the feasible region. And 131 SQP-runs converged to a local minimum, fulfilling the KKT-condition with high accuracy:

$$|\nabla f + \Sigma \lambda_i \nabla g_i| \leq 10^{-6}. \quad (36)$$

In the subset of the converged SQP-runs, 96 times the algorithm converged to an identical local minimum with $\tau^* = 0.3770$ s as presented in Figure 4. The remaining 35 times the algorithm converged to "exotic" local minima with unacceptable high values of the goal function: $\tau^* > 3$ s. As such we can conclude that the optimal control problem ((25)–(30)) possesses several local minima, which is not surprising due to convex-concave nature of the problem. But the best local minima can be clearly identified as the "global" solution, and therefore we accept the local minimum presented in Figure 4 as the solution of ((25)–(30)).

Naturally, the equality constraints are all active, and therefore their corresponding Lagrange multipliers are positive. Beyond those, at the solution, in total 60 inequality constraints are active with positive Lagrange multipliers. From this, we can conclude, that we can accept the solution computed by the SQP-algorithm as the solution of the optimal control problem.

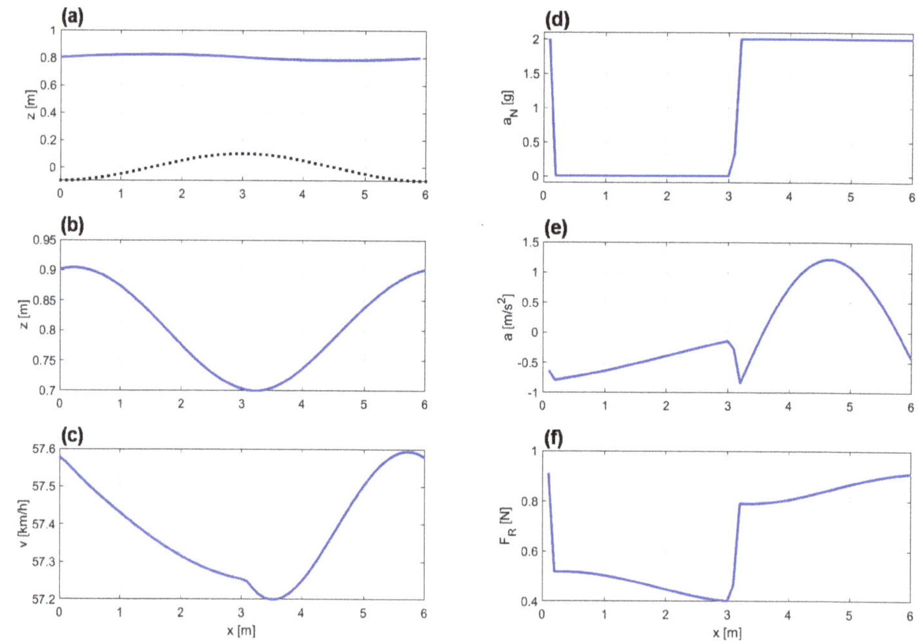

Figure 4. Optimal movement pattern, with $\tau^* = 0.3770$ s, for a track of length $L = 6.0$ m and a sinusoidal height profile with a difference of 0.2 m between the lowest and the highest point: (**a**) track profile (dotted line), trajectory of the athlete's COM (full line), (**b**) effective body height $h(t)$, (**c**) magnitude of the velocity of the COM $v(t)$, (**d**) magnitude of the normal acceleration $a_N(t)$, (**e**) magnitude of the acceleration of the COM $a(t)$, and (**f**) magnitude of the total friction force $F_R(t)$.

The calculations were done on a standard PC. We used the SQP-algorithm proposed by Han & Powell [12] as integrated in Matlab 2020a. Depending on the quality of the initial value for the SQP-algorithm, approximately 200 up to 1000 iterations were required to solve the optimal control problem with the required accuracy, which is a totally acceptable value for such a problem. Computation time on a standard time were approximately 1 min up to 10 min.

The athlete moves the COM in vertical direction with a minimal effective body height slightly after the track's peak, and a maximal effective body height slightly after the track has reached its valley (Figure 4a,b). Note that the athlete does not stretch his or her body to the maximum, but only to a level slightly more than the track's bump height. Quite obviously, it is advantageous for the athlete to keep the COM's movement stable on the same height, while major movement is done by the legs, in accordance to what is observed in sports. The magnitude of athlete's velocity varies between 57.2 km/h and 57.6 km, where the maximal velocity is reached slightly before the track's valley and and minimal velocity is slightly after the track's peak (Figure 4c).

In our eyes, the data presented in Figure 4d deliver a good insight into the optimal movement pattern, since the normal force $a_N(t)$ switches between 0 and $2g$, which is the maximal acceleration to be applied by the athlete. As such, the optimal movement pattern can be understood as a jumping-flying-landing-motion. The jump is performed in such a way that the athlete leaves the track exactly at the valley and performs a parabola-flight over the hill. At the downhill-side of the hill, the athlete uses its maximal body force for the landing and the preparation for the next jump. Loosely speaking, the athlete avoids driving uphill by jumping. Instead, the athlete only drives downhill. This picture delivers an intuitive and yet scientifically correct understanding of the energy pumping on a one-dimensional track. In Figure 4f, we see that the friction force acting on the athlete is largely affected by the jumping and landing. In the flying-phase the friction force is considerably smaller than in the landing phase, since the there is no sliding or rolling friction in athlete's flying-phase.

3.2. Optimal Movement Patterns under Variation of Athlete's Maximal Body Force and Track Height

In this section, we investigate (a) the effect of maximal vertical body force, which the athlete can apply for energy pumping, and (b) the effect of the track height, on the horizontal velocity. The parameters are chosen as in the previous section.

For the energy pumping on a track with sinusoidal height profile and circular conditions we find that there is a lower limit of athlete's maximal body force of approx. (1.2–1.3) g, below that the NLP does not have a solution, and accordingly energy pumping cannot be performed. For maximal body forces above this limit, the velocity increases steadily as shown in Figure 5 (right). Also, we see a pronounced influence of the track height on the speed. For track heights lower than $(h^{max} - h^{min})$, the velocity increases as the track height increases. In these cases the athlete is able to jump over the roller. But as the track height is larger than $(h^{max} - h^{min})$, the velocity decreases as the athlete is not able anymore to jump over the rollers.

In Figure 5 (left), we present the optimal body force as applied by the athlete for 4 distinct values for energy pumping on a track with track height 0.20 m. We see that the quality of the optimal movement remains untouched: In all cases we have a switch between a flying phase and a force phase. And in all cases, the athlete applies his or her maximal body force, but for a high maximal body force, the flying phase can be long, and the force phase is short. On the other hand, for a small maximal body force, the athlete can only perform a movement with a short flying phase, and the force phase needs to be long. Clearly, the exact timing of the flying and the force phase is fundamental to achieve optimal energy pumping.

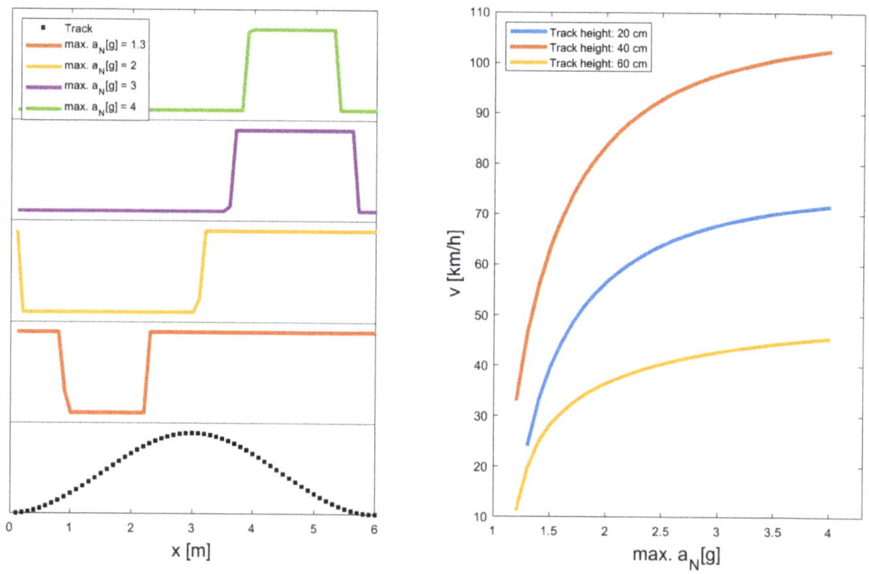

Figure 5. Variation of the maximal body force of the athlete on a track of length $L = 6.0$ m and a sinusoidal height profile with three different heights (0.20 m; 0.40 m; 0.60 m) between the lowest and the highest point: (**left**) Sinusoidal track profile with 0.20 m track height (dotted black line), dynamics of the body force as applied by athlete for different maximal body forces for the track height 0.2 m only. The graphics is scaled and shifted in vertical direction for clarity. (**right**) Variation of the athlete's medium velocity in x-direction under variation of the maximal body force for three different track heights (0.20 m; 0.40 m; 0.60 m).

3.3. *Qualitative Comparison of Optimal Movement Pattern with Data Derived from Video-Recorded BMX-Athlete*

Finally, we present a qualitative comparison of (a) the results achieved for the optimal movement pattern of the COM, to (b) a single elite athlete's movement pattern as taken from video from a public source [22]. This video does not allow for a quantitative comparison. The video only allows for a simple qualitative comparison. It is subject to future research to perform more precise and controlled measurements of movement patterns of a group of elite athletes for a more fundamental comparison.

For this, we applied manual tagging of the athlete's bike and athlete's body segments (head, shoulder, arms, upper leg, lower leg) for every single picture of the video sequence (see Figure 6). Then, the COM of the athlete and the bike was computed on the basis of standard weight distributions for human bodies. With this, we were able to compute the movement of the COM in the 2-dimensional space and derive COM's distance to the track.

Figure 6. Extraction of COM's movement pattern from a video recording of an BMX elite athlete during energy pumping. Left column: snapshots of athlete's movement. Right column: Tagging of athlete's body segments, the bike, and the computed COM (pink cross).

As presented in Figure 7, the movement patterns are similar but not identical. They share common features: jump phase in the uphill section of the bump, and landing phase in the downhill section of the bump. The maximal stretching of the body is observed in both cases slightly before the maximum of the bump.

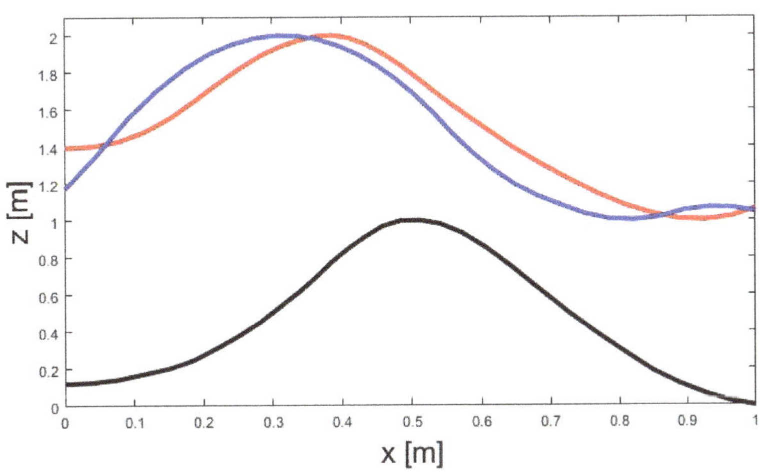

Figure 7. Movement patterns for energy pumping on a single bump (black line): movement of the COM as extracted from the video sequence (red line), and optimal movement pattern computed by solving the discrete optimal control problem.

4. Discussion

4.1. General

We present our results as a first step towards a full understanding of optimal movement patterns for energy pumping. Clearly, there is a tremendous space for future research and important improvements. In our eyes, the following future research directions seem to be most promising, and interesting:

1. As the inverted pendulum model presented in this paper is highly simplified, we propose to increase the accuracy of the model by integrating more degrees of freedom into the model (modeling of athlete's joints, modeling of athlete's equipment). With this, the physical model of the athlete's movement will be closer to reality, and therefore it is more likely that concise and accurate conclusions can be drawn from the computed optimal movement patterns to support athletes, and contribute to sport science. For instance, we expect results, which are closer to reality, if the athlete does not only have a single isolated contact point (such as in the inverted pendulum model), but instead has multiple contact points (as it is in biking), or a line of contact points (as it is skiing).
2. Perform high-quality measurements of movement patterns of athletes performing energy pumping in such different disciplines, such as BMX-racing, or ski-cross. On this basis, it would be highly interesting to be able to compare movement patterns of athletes from different disciplines to the results of the optimal movement patterns as presented in this paper.

4.2. Learnings for Training and Competition in Sports

From our study we can draw some conclusions for athletes and coaches engaged in sports in which energy pumping is an success factor (like BMX racing, or ski cross):

1. The results fully agree with athlete's experience that (a) leg force should be applied before the bump to avoid uphill movement, and (b) in the landing or compression phase, leg force is needed for landing.
2. In any case, if facing high or low bumps, the athlete should use his or her maximal leg force, since the horizontal acceleration depends on the maximal leg force applied.
3. The correct and exact timing of jump and landing phase is crucial for the energy pumping. Obviously, as horizontal velocities increase it is increasingly difficult for the athlete to achieve the correct timing of the jumping and landing phase.
4. The optimal path can be used for evidence-based coaching.
5. As a result of our study, the major advice for athletes could be phrased in simple words as: Avoid riding uphill, it's better to jump over the hill. Generally when driving downhill try to push your center of mass upwards.

4.3. Limitations

As this study is theoretical in nature, its focus lies on defining a proper dynamical model, defining a proper continuous and discrete optimal control problem, and solving the latter numerically according to the state-of-the-art. Therefore, it is important for us to highlight the limitations of our work:

1. In this study there was no quantitative validation. In the future a proper quantitative validation is required based on subject-specific scaling of the model and motion analysis on the field.
2. While the simple inverted-pendulum model turns out to be an adequate model to understand energy pumping in a generic way, it might be oversimplified to deliver a detailed description of the movement. As such, in the future, we expect more complex models will be required for a detailed description of optimal movement patterns in energy pumping.
3. As the inverted-pendulum model possesses a single contact point between the athlete and the ground, it might be oversimplified for a detailed description of cycling (with

2 contact points), and skiing (with a extended line of contact). Therefore, in the future, to describe the optimal movement patterns in those sports in detail an extension of the inverted-pendulum model is required.

5. Conclusions

The inverted pendulum model, which is already highly simplified, in combination with the numerical solution of an optimal control problem turn out to be valuable tools to increase the fundamental understanding of the effect of energy pumping. Although, more work has to be done to increase the accuracy of the model on one hand, and on the other hand to provide a more exact and broader basis of experimental measurements, our research allows already for practically relevant conclusions, and even more importantly, provides a guide to the eye for the future research. The authors find it especially promising to investigate in the future optimal movement patterns in other sports like gymnastics, skiing, or track and field.

Author Contributions: Conceptualization, M.J.B. and S.L.; methodology, M.J.B.; software, M.J.B. and M.L.; validation, S.L., M.G. and D.G.; formal analysis, M.J.B.; investigation, M.L.; data curation, M.L.; writing—original draft preparation, M.J.B.; writing—review and editing, S.L. and M.G.; visualization, M.L.; supervision, M.J.B. All authors have read and agreed to the published version of the manuscript.

Funding: This research received no external funding.

Institutional Review Board Statement: Ethical review and approval were waived for this study due to the fact that only an anonymized video out of a normal training setting was used. The one participant agreed to be part of the study.

Informed Consent Statement: Informed consent was obtained from all subjects involved in the study.

Data Availability Statement: Not applicable.

Conflicts of Interest: The authors declare no conflict of interest.

References

1. Mateo, M.; Blasco-Lafarga, C.; Zabala, M. Pedaling power and speed production vs. technical factors and track difficulty in bicycle motocross cycling. *J. Strength Cond. Res.* **2011**, *25*, 3248–3256. https://doi.org/10.1519/JSC.0b013e3181f90847.
2. Schärer, C.; Gross, M.; Lorenzetti, S.; Hübner, K. Energy Transformation on Vault in Elite Artistic Gymnastics: Comparisons between Simple and Difficult Tsukahara and Yurchenko Vaults. *Appl. Sci.* **2021**, *11*, 9484. https://doi.org/10.3390/app11209484.
3. Ouadahia, N.; Ababoua, A.; Ababoua, N.; Larbib, M.A. Windsurf Ergometer for Sail Pumping Analysis and Mechanical Power Measurement. *Procedia Eng.* **2014**, *72*, 249–254. https://doi.org/10.1016/j.proeng.2014.06.045.
4. Piccoli, B.; Kulkarni, J. Pumping a swing by standing and squatting: Do children pump time optimally? *IEEE Control Syst. Mag.* **2005**, *25*, 48–56. https://doi.org/10.1109/MCS.2005.1499390.
5. Curry, S.M. How children swing. *Am. J. Phys.* **1976**, *44*, 924–934. https://doi.org/10.1119/1.10230.
6. Nitschke, M.; Dorschky1, E.; Heinrich, D.; Schlarb, H.; Eskofierl, B.M.; Koelewijn1, A.D.; van den Bogert, A.J. Efficient trajectory optimization for curved running using a 3D musculoskeletal model with implicit dynamics. *Nature* **2020**, *10*, 874–884. https://doi.org/10.1038/s41598-020-73856-w.
7. Nguyen, V.Q.; Johnson, R.T.; Sup, F.C.; Umberger, B.R. Bilevel Optimization for Cost Function Determination in Dynamic Simulation of Human Gait. *IEEE Trans. Neural Syst. Rehabil. Eng.* **2019**, *27*, 1426–1435. https://doi.org/10.1109/TNSRE.2019.2922942.
8. Umberger, B.R. Stance and swing phase costs in human walking. *J. R. Soc. Interface* **2010**, *7*, 1329–1340. https://doi.org/10.1098/rsif.2010.0084.
9. Umberger, B.R.; Miller, R.H. Optimal Control Modeling of Human Movement. In *Handbook of Human Motion*; Müller, B., Wolf, S.I., Eds.; Springer: Berlin/Heidelberg, Germany, 2017; pp. 1–22. https://doi.org/0.1007/978-3-319-30808-1_177-1.
10. Liberzon, D. *Calculus of Variations and Optimal Control Theory*; Princeton University Press: Princeton, NJ, USA, 2012. https://doi.org/10.1515/9781400842643.
11. Nocedal, J.; Wright, S.J. *Numerical Optimization*, 2nd ed.; Springer Science+Business Media: Berlin, Germany, 2006. https://doi.org/10.1007/978-0-387-40065-5.
12. Han, S.P. A globally convergent method for nonlinear programming. *J. Optim. Theory Appl.* **1977**, *22*, 297–309. https://doi.org/10.1007/BF00932858.
13. Powell, M.J.D.; Watson, G.A. A fast algorithm for nonlinearly constrained optimization calculations. In Proceedings of the Biennial Conference, Dundee, UK, 28 June–1 July 1977; pp. 144–157. https://doi.org/10.1007/BFb0067703.

14. Schittkowski, K. NLPQL: A Fortran subroutine for solving constrained nonlinear programming problems. *Ann. Oper. Res.* **1986**, *5*, 485–500. https://doi.org/10.1007/BF02022087.
15. van de Braak, G.; Bünner, M.J.; Schittkowski, K. Optimal Design of Electronic Components by Mixed-Integer Nonlinear Programming. *Optim. Eng.* **2004**, *5*, 271. https://doi.org/10.1023/B:OPTE.0000038887.72677.3e.
16. F. Gfeller, L.V.; Schramm, M.; Bünner, M. Optimization of the Lift Curve of Airfoils with Special Emphasis on the Improvement of the Airfoil's Stall Behavior. In Proceedings of the 6th Symposium on OpenFOAM in Wind Energy, Gotland, Sweden, 13 June 2018.
17. Nagel, M.G. Numerische Optimierung Dreidimensional Parametrisierter Turbinenschaufeln mit Umfangsunsymmetrischen Plattformen–Entwicklung, Anwendung, Validierung. Ph. D. Thesis, Universität der Bundeswehr München, Neubiberg, Germany, 2004.
18. Achtziger, W. Truss topology optimization including bar properties different for tension and compression. *Struct. Optim.* **1996**, *12*, 63–74. https://doi.org/10.1007/BF01270445.
19. Bünner, M.J. The Mathematics of Optimal Products. In Proceedings of the NAFEMS Conference, Wiesbaden, Germany, 10–11 November 2014.
20. Willems, P.A.; Cavagna, G.A.; Heglund, N.C. External, Internal and Total Work In Human Locomotion. *J. Exp. Biol.* **1995**, *198*, 379–393.
21. Peyré-Tartaruga, L.A.; Dewolf, A.H.; di Prampero, P.E.; Fábrica, G.; Malatesta, D.; Minetti, A.E.; Monte, A.; Pavei, G.; Silva-Pereyra, V.; Willems, P.A.; et al. Mechanical work as a (key) determinant of energy cost in human locomotion: Recent findings and future directions. *Exp. Physiol.* **2020**, *106*, 1897–1908. https://doi.org/10.1113/EP089313.
22. Pushen. Available online: https://www.youtube.com/watch?v=Y2-tjzYYWHA (accessed on 1 December 2022).

sports

Article

Effectiveness of Different Modalities of Remote Online Training in Young Healthy Males

Michael Daveri [1], Andrea Fusco [2], Cristina Cortis [2] and Gabriele Mascherini [1,*]

[1] Department of Experimental and Clinical Medicine, University of Florence, 50134 Florence, Italy
[2] Department of Human Sciences, Society and Health, University of Cassino and Lazio Meridionale, 03043 Cassino, Italy
* Correspondence: gabriele.mascherini@unifi.it

Abstract: Since 2020 there has been an increase in demand for home workouts. Therefore, different ways of delivering distance training have been proposed to promote "stay active at home." This study aimed to compare the effectiveness of three different training programs consisting of a total of 15 workouts (three sessions per week): supervised livestreaming (LS), unsupervised following a video recording (VR), and unsupervised following a written program (WP). Changes in anthropometric and cardiovascular variables, muscle fitness, and physical activity levels were evaluated. To provide a meaningful analysis for significant comparisons between small groups, mean differences (Δ), 95% confidence interval (95% C.I.), and Cohen's effect sizes (E.S.) were also calculated. The three training modalities increased physical activity levels, with an adherence rate of LS = 93.3%, VR = 86%, and WP = 74%. Although there was no reduction in body weight, waist circumference decreased by 1.3 cm (95% C.I. = −2.1, −0.5; E.S. = 0.170; $p < 0.004$). Furthermore, where LS, VR, and WP resulted in improvements in muscle fitness, only LS showed changes in cardiovascular variables, such as resting heart rate (Δ = −7.3 bpm; 95% C.I. = −11.9, −2.7; E.S. = 1.296; $p < 0.001$) and Ruffier's index (Δ = −2.1bpm; 95% C.I. = −3.5, −0.8; E.S. 1.099; $p < 0.001$). Remote online training proved its effectiveness over a short period of time. However, supervised training proved to be the most effective, highlighting the importance of an experienced trainer.

Keywords: livestreaming exercise; on-demand exercise; home exercise; supervised exercise

check for updates

Citation: Daveri, M.; Fusco, A.; Cortis, C.; Mascherini, G. Effectiveness of Different Modalities of Remote Online Training in Young Healthy Males. *Sports* **2022**, *10*, 170. https://doi.org/10.3390/sports10110170

Academic Editors: Flávio Antônio de Souza Castro, Rui Miguel Simões de Azevedo and Rodrigo Zacca

Received: 13 September 2022
Accepted: 31 October 2022
Published: 2 November 2022

1. Introduction

On 30 January 2020, the World Health Organization (WHO) declared the coronavirus pandemic a global emergency. In Italy on 9 March 2020, non-essential and strategic activities such as gyms, swimming pools, wellness centers, and clubs were suspended [1]. The coronavirus outbreak brought enormous challenges for the fitness training sector, with an increased use of technology and the development of new training strategies [2]. As the pandemic progresses, most activities have been moved online, including medical screening and participation in physical activity [2].

Performing sedentary behaviors for at least seven hours a day carries an independent risk factor for numerous chronic diseases and overall mortality [3]. Therefore, the promotion of physical activity is an action declared by the WHO, aiming for an active lifestyle that favors the benefits of physical training, thus reducing the consequences that physical inactivity entails [4].

During the pandemic, there was an increase in online searches for "home workout" [5]. Therefore, the use of the internet has been exacerbated as well as the use of technology, as individuals tried to stay active through applications, websites, videos, and social media, engaging in remote online workouts [6].

Previous studies showed some advantages regarding online training, showing how it can bring partial benefits compared to face-to-face training [6]. Other recent research shows the feasibility of online remote training for diseased populations, such as cancer and multiple sclerosis patients [7,8].

It seems that the presence of the instructor in the face-to-face mode brings advantages in terms of motivation and attention, which then translate into better fitness parameters [6]. However, the instructor's supporting role during online training programs has not yet been studied. The hypothesis of this study is that the supervision of the instructor can provide greater improvements, even in remote online training.

It is conceivable that the use of technology in remote training could continue even after the end of the coronavirus pandemic; therefore, evaluating the effectiveness of different approaches can be helpful to choose the most effective solution to engage people in physical activity [9].

Considering the advantage of online training during the pandemic where social distancing is required, this study aims to investigate the effects of three different ways of remote online training on anthropometric and cardiovascular variables, muscle fitness, and physical activity levels, by evaluating their feasibility and effectiveness.

2. Materials and Methods

2.1. Participants and Study Design

The recruitment process involved the dissemination of the initiative among university students. Those who: (1) gave their availability, (2) guaranteed the achievement of the sample size, (3) met the inclusion criteria, and (4) signed informed consent were enrolled in the study. Twenty-one healthy males (23.1 ± 1.5 years; height 180 ± 0.1 cm; weight 73.4 ± 8.7 kg) were enrolled in this study. Inclusion criteria were: aged 18–25 years, no contraindications to exercise, access to the internet, and availability for approximately 2 h per day to spend in the training sessions.

Exclusion criteria were: aged under 18 or over 25 years, no access to the web platform used, recent injuries, practicing sports at competitive levels, unavailable during the 2-h training sessions.

The study was undertaken in accordance with the ethical standards laid down in the 1975 Declaration of Helsinki. The Institutional Review Board of the Department of Human Sciences, Society, and Health of the University of Cassino and Lazio Meridionale approved this study (approval No.: 9947; date: 14 April 2021).

Eligible subjects were divided into 3 groups (n = 7 for each group) following the model for randomized studies with parallel groups:

1. Livestream group (LS): the subjects received the instructions directly from the experienced trainer via livestream on a web platform. The experienced trainer explained each exercise; his direct intervention showed the correct execution, corrected any errors, motivated the group, and supervised the execution in terms of sets, repetitions, rest, and range of motion.
2. Video recording group (VR): the subjects received the instructions from a video recording uploaded on a web platform. The videos showed the different training sessions in terms of exercises, the correct execution, sets, repetitions, and rests. The experienced trainer did not participate live during the training sessions, and each subject had to train individually at any time of the day. However, the experienced trainer was fully available (phone call, instant message, etc) to respond to any requests from participants.
3. Written program group (WP): the subjects received instructions from a written training program. The program was first explained in detail, then sent and made available on smartphones for greater daily availability. The program showed the exercises through photos and descriptions of the execution and series, repetitions, and rest. Similar to the VR group, the experienced trainer did not participate directly during the training sessions, although they were available upon request.

The evening before the training day, a reminder was sent to all participants when the training session was uploaded to the group platform (link for LS, video recording for VR, and program sent by phone message for WP). In addition, to verify adherence to the training program, members of the VR and WP groups had to send a statement at the end of each completed workout, while for the LS group, the experienced trainer recorded the participation.

2.2. Procedure

The effectiveness of the 3 training programs was evaluated by comparing the pre-post training results of anthropometric evaluations, flexibility and balance, muscle strength, cardiovascular parameters, and weekly physical activity level. The experienced trainer performed assessments at each subject's home to ensure that the test was performed correctly. Therefore, 3 attempts were granted for each test and the best result was considered for further analysis [10].

2.2.1. Anthropometric Variables

Weight was measured to the nearest 0.1 kg and height to the nearest 0.1 cm (Seca 786 GmbH and Co. Hamburg, Germany). Body mass index (BMI) was calculated as the body mass divided by height squared (kg/m^2). Waist circumference measures were made with a tape metric (Holtain Limited, Crymych, UK, 1.5 m Flexible Tape) at the narrowest level, or if this was not apparent, at the midpoint between the lowest rib and the top of the hip bone (iliac crest) [11].

2.2.2. Muscle Fitness

Muscle fitness was assessed by testing for: (1) flexibility, (2) joint mobility, (3) balance, and (4) muscle strength [12].

1. The sit and reach (SR) test was used as a measure of the posterior extensibility of the body [13]. Measurements were taken as the distance from fingertips to toes during trunk flexion.

2. Shoulder flexion and extension tests measure the angles of flexion and extension of the shoulder and evaluate the possible retraction of the shoulder girdle muscles. The shoulder internal rotation test measures the angle of internal arm rotation from 90° abduction and 90° forearm flexion and allows evaluation of the external shoulder rotation functionality [14].

3. The one leg test measures the ability to balance in a single stance with eyes closed. This enables the evaluation of body sensory management [15].

4. The hand grip test estimates the overall static strength of the upper limb. The subjects were in a sitting position with the shoulders adducted and rotated in a neutral position, the elbow flexed at 90°, and the forearm in a neutral position. Three tests were carried out on each side [16]. The maximal push-up test was administered to evaluate upper-body muscle endurance by repetition until muscle failure or incorrect execution. For a correct push up, the subject had to keep his back straight, and at each stretch up the elbows had to lock [17]. Finally, the maximal plank test was administered to evaluate core muscle strength. Subjects had to stand prone with shoulders and elbows flexed at 90°, maintaining a straight line from head to toe without lowering the hips, keeping the neck in a neutral position [18]. Subjects were required to maintain the position for as long as possible and the time was recorded.

2.2.3. Cardiovascular Variables

When the participants were seated and relaxed for at least 3 min, the resting heart rate was measured by pulse oximeter (CocoBear, Pulse Oximeter PR-10). Resting heart rate is an independent cardiovascular risk factor, and it could help to identify potential health problems [19]. For effort tolerance, the Ruffier test was used. This allows the evaluation of the subject's recovery capacity from an effort. Each subject was asked to perform 30 squats in 45 s, followed by heart rate recording; after 1 min of rest, heart rate was measured again and the Ruffier index was estimated [20].

2.2.4. Physical Activity Level

The international physical activity questionnaire (IPAQ) provided standard instruments that could be used to gain internationally comparable data on health-related physical activity [21]. The questionnaire, which queries time spent physically active in the last 7 days, comprises of 4 sections:

1. Vigorous physical activities: minutes per week. This refers to activities of hard physical effort which make breathing much harder than usual.
2. Moderate activities, but not walking: minutes per week. These activities refer to actions of moderate physical effort which make breathing somewhat harder than usual.
3. Walking: minutes per week and number of daily steps. This includes at work and home, walking to travel from place to place, and any other walking undertaken solely for recreation, sport, exercise, or leisure.
4. Time spent sitting, including reading, watching television, studying, and playing video games: minutes per day.

2.3. Training Program

The training program lasted 5 weeks with a frequency of 3 sessions per week (Monday, Wednesday, Friday) for a total of 15 sessions.

Each training session included 3 parts: warm-up with flexibility and range of motion exercises (20 min), whole-body muscle strengthening (45 min), and cool-down with static muscle stretching (15 min). The whole-body muscle strengthening part was based on exercises individualized to the participants' ability, ensuring that participants could perform those exercises at home independently. Therefore, this mainly included bodyweight exercises, such as push-ups, squats, lunges, scapular adductions and planks, as these do not require special equipment. The program was based on the methodology training fundamental principles; the routine had a progressive load increase each weeks, had training load variety through different kinds of exercises, and training load specificity to improve participants' strength levels and improve posture [22]. In order to verify the effectiveness of the 3 training programs, all 3 groups followed the same training program and the same progression of the load. During the first week, each group performed 3 sets of 12 repetitions with 60 s of recovery between sets for each exercise. During the second and third weeks, the sets were increased to 5, and in the fourth and fifth weeks the repetitions were increased to 15. The main muscles involved in the training program were the pectoral, back, thigh, shoulder, and core muscles, such as the pectoralis, latissimus dorsi, quadriceps, hamstrings, gluteus, deltoids, and abdominal.

2.4. Statistical Analysis

Stata statistical software version 14.2 (Stata-Corp, College Station, TX, USA) was used for statistical analysis. All variables were calculated by mean, standard deviation (SD), mean difference between pre-post (Δ), and 95% confidence intervals (95% C.I.). The Cohen's d effect size (E.S.) was calculated to determine the magnitude of effect. The E.S. was assessed using the following criteria: small < 0.20, medium < 0.50, and large < 0.80. Multilevel regression models (or hierarchical linear models) were performed to examine the effects of different training modalities on the subject's strength, mobility, and cardiorespiratory fitness performances. Subjects were considered the random effect, whereas the training modalities and testing time (pre vs. post) were treated as fixed effects. The models were fitted using the maximum residual likelihood for accounting for the small sample. The repeated measures ANOVA method was used for computing the degrees of freedom of a t distribution, as subjects were tested before and after the training protocols. Subsequently, the contrast method was used to test whether the dependent variable (i.e., the Ruffier test and push-up test) meant each training modality and time were identical. The contrast method tests included ANOVA-style tests of the main effects used to make comparisons against the reference categories (i.e., pre vs. post; LS vs. VR vs. WP). The statistical significance of the main effect on the whole sample analysis was set at $p < 0.05$. Subsequently, Bonferroni post-hoc tests were used for multiple-comparison adjustments across all terms where significant main effects and interactions were found. After Bonferroni correction, statistical significance was set at $p < 0.003$.

3. Results

The remote online exercise program adherence rate in the three groups was LS = 93.3%, VR = 86%, and WP = 74%.

3.1. Changes after Remote Training on the Whole Sample

The effects of remote training on the whole sample are shown in Table 1.

Table 1. Changes after remote training in anthropometric and cardiovascular variables, in muscle fitness, and physical activity level in the whole study sample of 21 subjects. Data are expressed as mean and S.D. for the results obtained before (PRE) and after (POST) training. Δ variation between baseline and post; C.I. (95%) = 95% confidence intervals; E.S. = effect size.

		PRE	POST	Δ	C.I. (95%)	E.S.	*p* Value
Anthropometric variables	Weight (kg)	73.4±8.7	73.2±8.6	−0.29	−1.4 | 0.8	0.023	0.579
	BMI (Kg/m^2)	22.6 ± 2.3	22.5 ± 2.2	−0.10	−0.4 | 0.2	0.044	0.539
	Waist Circumference (cm)	83.6 ± 6.0	82.3 ± 5.7	−1.29	−2.1 | −0.5	0.170	0.004
Muscle fitness	Sit and Reach (cm)	26.1 ± 8.6	29.0 ± 7.2	3.4	0.7 | 6.2	0.365	0.018
	Shoulder Flexion (°)	174.2 ± 7.0	185.5 ± 5.0	11.0	5.6 | 16.6	1.858	0.000
	Shoulder Extension (°)	60.5 ± 9.7	66.8 ± 8.8	4.6	−0.9 | 10.2	0.680	0.097
	Shoulder Internal Rotation (°)	56.4 ± 16.9	69.3 ± 10.1	15.0	5.7 | 24.3	0.926	0.003
	One Leg Test (s)	23.9 ± 18.0	55.5 ± 47.4	32.8	3.6 | 61.9	0.881	0.030
	Handgrip (kg)	88.7 ± 21.3	95.6 ± 19.3	5.9	0.4 | 11.5	0.339	0.036
	Maximal Push-up (rep.)	26.3 ± 10.1	34.0 ± 7.9	8.7	5.0 | 12.4	0.849	0.000
	Maximal Plank (s)	90.8 ± 26.7	117.4 ± 30.4	29.1	17.2 | 41.1	0.930	0.000
Cardiovascular variables	Rest HR (bpm)	69.0 ± 7.2	66.3 ± 4.1	−7.3	−10.6 | −3.9	0.461	0.000
	Ruffier Index (bpm)	9.0 ± 2.2	7.9 ± 1.6	−2.1	−3.1 | −1.2	0.572	0.000
Physical activity level	IPAQ (METs)	1639.6 ± 1531.6	2991.4 ± 1715.3	1304.3	360.4 | 2248.2	0.831	0.009

3.1.1. Anthropometric Variables

Based on the anthropometric parameters, the sample included normal weight subjects within the normal range with waist circumference. On average, there was no significant weight loss (pre 73.4 ± 8.7 kg, post 73.2 ± 8.6 kg), and therefore the BMI did not decrease either (pre 22.6 ± 2.3 kg/m^2; post 22.5 ± 2.2 kg/m^2). On the other hand, the waist circumference shows a significant reduction (pre 83.6 ± 6.0 cm; post 82.3 ± 5.7 cm; $p < 0.004$).

3.1.2. Muscle Fitness

Muscle fitness parameters show significant increases in the muscle strength component evaluated in the push-up (pre 26.3 ± 10.1 rep.; post 34.0 ± 7.9 rep.; $p < 0.001$), plank (pre 90.8 ± 26.7 s; post 117.4 ± 30.4 s; $p < 0.001$), and handgrip (pre 88.7 ± 21.3 kg; post 95.6 ± 19.3 kg; $p = 0.036$) tests. For the joint mobility component of muscle fitness, only shoulder extension did not show significant differences (pre 60.5° ± 9.7; post 66.8° ± 8.8). However, the other joint mobility assessments show significant increases in shoulder flexion (pre 174.2° ± 7.0; post 185.5° ± 5.0; $p < 0.001$) and internal rotation (pre 56.4° ± 16.9; post 69.3° ± 10.1; $p = 0.003$). Flexibility and balance show improvements in the sit and reach test (pre 26.1 ± 8.6 cm; post 29.0 ± 7.2 cm; $p = 0.018$) and one leg test (pre 23.9 ± 18.0 s; post 55.5 ± 47.4 s; $p = 0.03$).

3.1.3. Cardiovascular Variables and Physical Activity Level

Cardiovascular variables showed a significant reduction in resting heart rate (baseline 69.0 ± 7.2 bpm; post 66.3 ± 4.1 bpm) and Ruffier's index (baseline 9.0 ± 2.2 bpm; post

7.9 ± 1.6 bpm). Physical activity levels also increased significantly (baseline 1639.6 ± 1531.6 METs; post 2991.4 ± 1715.3 METs).

3.2. Changes after Remote Training According to the Division into Groups LS, VR, and WP

For the variables that showed significant differences in the whole group analysis, an analysis was performed to verify the effectiveness of remote training for each delivery modality used (Tables 2–4).

Table 2. Effectiveness of the livestream (LS) delivery method for remote training. Data are expressed as mean and S.D. for the results obtained before (PRE) and after (POST) training. Δ variation between baseline and post. C.I. (95%) = 95% confidence intervals; E.S. = effect size.

		LS					
		PRE	POST	Δ	C.I. (95%)	E.S.	*p* Value
Muscle fitness	Shoulder Flexion (°)	175.4 ± 6.8	186.4 ± 6.4	11.1	3.4 | 18.7	1.666	0.000
	Handgrip (kg)	85.9 ± 22.0	91.8 ± 23.3	5.9	−1.8 | 13.7	0.260	0.358
	Maximal Push-up (rep.)	23.1 ± 12.1	31.9 ± 8.7	8.7	3.6 | 13.9	0.835	0.000
	Maximal Plank (s)	85.4 ± 27.3	114.6 ± 26.3	29.1	12.4 | 45.9	1.089	0.000
Cardiovascular variables	Rest HR (bpm)	73.4 ± 7.0	66.1 ± 3.8	−7.3	−11.9 | −2.7	1.296	0.000
	Ruffier Index (bpm)	9.8 ± 2.1	7.7 ± 1.7	−2.1	−3.5 | −0.8	1.099	0.000

Table 3. Effectiveness of the video recording (VR) delivery method for remote training. Data are expressed as mean and S.D. for the results obtained before (PRE) and after (POST) training. Δ variation between baseline and post. C.I. (95%) = 95% confidence intervals; E.S. = effect size.

		VR					
		PRE	POST	Δ	C.I. (95%)	E.S.	*p* Value
Muscle fitness	Shoulder Flexion (°)	173.9 ± 8.3	185.7 ± 4.5	11.7	4.1 | 19.4	1.767	0.000
	Handgrip (kg)	90.8 ± 23.6	101.0 ± 17.7	10.16	2.4 | 17.9	0.489	0.002
	Maximal Push-up (rep.)	25.4 ± 10.5	33.0 ± 8.1	7.57	2.4 | 12.7	0.810	0.000
	Maximal Plank (s)	89.6± 27.1	111.1 ± 24.9	21.57	4.8 | 38.3	0.826	0.002
Cardiovascular variables	Rest HR (bpm)	68.6 ± 8.6	65.7 ± 5.7	−2.9	−7.5 | 1.8	0.397	1.000
	Ruffier Index (bpm)	9.8 ± 1.4	8.2 ± 1.2	−1.6	−2.3 | −0.3	1.227	0.006

No differences in delivery modalities of remote training were observed in:

- waist circumference (LS: baseline 85.5 ± 3.6 cm, post 84.2 ± 3.5 cm, p = 0.012; VR: baseline 85.6 ± 7.1 cm, post 84.6 ± 6.9 cm, p = 0.080; WP: baseline 79.6 ± 5.5 cm, post 79.1 ± 5.3 cm, p = 1)
- sit and reach (LS: baseline 25.7 ± 9.5 cm, post 29.1 ± 6.8 cm, p = 0.134; VR: baseline 23.3 ± 10.3 cm, post 26.9 ± 9.4 cm, p = 0.097; WP: baseline 29.4 ± 5.3 cm, post 31.1 ± 5.1 cm, p = 1)
- shoulder internal rotation (LS: baseline 56.4° ± 17.9, post 71.4° ± 10.1, p = 0.011; VR: baseline 60.0° ± 19.1, post 69.6° ± 10.4, p = 0.445; WP: baseline 52.9° ± 15.6, post 66.8° ± 10.8, p = 0.025)
- one leg test (LS: baseline 16.2 ± 9.5 s, post 49.0 ± 39.4 s, p = 0.274; VR: baseline 24.3 ± 19.0 s, post 43.9 ± 29.1 s, p = 1; WP: baseline 31.2 ± 22.3 s, post 73.7 ± 67.2 s, p = 0.033)
- IPAQ (LS: baseline 1427.1 ± 1849.6 METs, post 2731.4 ± 1966.9 METs, p = 0.055; VR: baseline 1695.7 ± 1410.1 METs, post 3212.1 ± 1972.5 METs, p = 0.011; WP: baseline 1796.1 ± 1525.0 METs, post 3030.7 ± 1382.0 METs; p = 0.09)

Table 4. Effectiveness of the written program (WP) delivery method for remote training. Data are expressed as mean and S.D. for the results obtained before (PRE) and after (POST) training. Δ variation between baseline and post. C.I. (95%) = 95% confidence intervals; E.S. = effect size.

		WP					
		PRE	POST	Δ	C.I. (95%)	E.S.	*p* Value
Muscle fitness	Shoulder Flexion (°)	173.2 ± 6.9	184.3 ± 4.5	11.1	3.4 | 18.7	1.905	0.000
	Handgrip (kg)	89.3 ± 21.4	94.0 ± 18.2	4.7	−3.1 | 12.4	0.236	1.000
	Maximal Push-up (rep.)	30.3 ± 7.3	37.0 ± 7.1	6.7	1.6 | 11.9	0.930	0.002
	Maximal Plank (s)	97.4 ± 28.3	126.6 ± 40.2	29.1	12.4 | 45.9	0.840	0.000
Cardiovascular variables	Rest HR (bpm)	65.0 ± 2.4	67.1 ± 3.0	2.14	−2.5 | 6.8	0.773	1.000
	Ruffier Index (bpm)	7.5 ± 2.4	7.9 ± 2.1	0.46	−0.9 | 1.8	0.177	1.000

4. Discussion

The main findings were that changes in anthropometric, cardiovascular, and muscle fitness variables occurred as a result of training programs performed: (1) under the supervision of an experienced trainer during a livestream, (2) without supervision following a video recording, or (3) without supervision following a written program. The results obtained from the whole sample show how remote online training effectively improves both muscle and cardiovascular fitness. In addition, the physical activity levels of all subjects also increased. While improvements in muscle fitness were reported in all three groups, improvements in cardiovascular variables were seen more so in the livestream group under the supervision of the experienced trainer. This aspect could be unexpected; from the supervision, we expect correct execution of the exercise and, therefore, a proper pattern of muscle contraction, resulting in increased muscle fitness. The results obtained on cardiovascular variables induce possible speculation regarding the role of supervised exercise, even remotely; correctly undertaking an exercise program results in further improvements. It is also likely better for management of training times in terms of the alternating effort–recovery of the entire workout. The lack of variation in anthropometric parameters, such as weight and BMI, could be caused by the lack of nutritional indications, the duration of the intervention, and the initial condition of the normal weight of the sample.

Previous studies on populations with chronic diseases [7,8] did not perform direct assessments as in the present study, but reported the rate of adherence (% sessions attended), retention (% participants who completed the intervention), and safety (number of events). The adherence rate of the LS group is in line with the study by Winters-Stone et al. [7]; however, a further comparison of the results of the fitness assessments obtained from this study is not possible.

The coronavirus pandemic has promoted technology to offer online remote health services. The planning and execution of medical examinations concerning sports and exercise medicine were also carried out remotely [23]. Other health activities, such as physiotherapy, also implemented services realized through online proposals [24,25].

Online remote physical training was used before the pandemic; in 2020, it ranked 26th in the worldwide survey of fitness trends. However, in 2021, it ranked 1st due to a shift in the fitness market from clubs to homes due to the pandemic. In 2022, it is 9th, as although coronavirus is still present, the need for greater social interaction likely resulted in this change [9]. However, some people may still want to maintain social distance during workouts. Therefore, determining which remote training modalities are the most effective could have an impact on the training program chosen and may also help to achieve the best adherence rate.

The indications provided by the fitness sector show an apparent effect of the pandemic on the European health and fitness industry [26]. People are aware that physical activity is healthy, but not everyone can or wants to go to the gym. Furthermore, online distance learning could be the first approach for new customers. Cronshaw's study [27] shows that those who have experienced the benefits of online distance training question whether they need to go back to a physical gym. Therefore, gym owners must develop another proposal, offering online classes to those who cannot or do not want to join the gym. Therefore, the market adapted by offering hybrid fitness services that have become a reality. However, this could compromise the ability to deliver such services by professionals accustomed to offering workouts in the traditional in-person modality. Considering long-term fitness consumption, the integrated digital–physical model can help different groups of people who face barriers when attending the gym.

This study has some limitations. Firstly, the reliability of the parameters regarding the self-reported adhesion of the unsupervised groups (VR and WP) must be considered. However, this aspect is entirely within the scope of the study: the increase in physical activity levels shows a continuous training engagement of the sample. Furthermore, compliance with the agreements between the experienced trainer and the subject plays a crucial role in the effectiveness of any training proposal. Secondly, the online remote training program did not include aerobic activity. The proposed program envisaged feasibility regardless of the use of specific tools. Different aerobic training tools (treadmill vs. stationary bike vs. rowing machine) may have been confounding factors in the results due to the different adaptations induced by different movement patterns. However, the results show that the proposed training program also managed to modify cardiovascular variables.

Practical Implications

- Remote online training has effectively become a way to deliver exercise. Therefore, the practitioner should consider this delivery model as an option for their clients.
- The three training methods proved to be effective. Therefore, the choice of delivery method should be based on the individual characteristics of the subjects and their preference.
- The livestreaming mode with the supervision of an experienced trainer was the most effective. Therefore, it should be preferred in the case of physical exercise on a single subject.

5. Conclusions

In summary, the present study results indicate that remote online training was effective in healthy young male adults for improving muscle and cardiovascular fitness. However, the livestreaming model with expert supervision appears to be the delivery modality that achieves the higher effects, especially on cardiovascular parameters.

Further studies in other subjects are needed in order to establish the effectiveness of remote online training in other study populations.

Author Contributions: Conceptualization, M.D. and G.M.; methodology, M.D.; software, A.F. and M.D.; validation, C.C. and G.M.; formal analysis, M.D. and A.F.; investigation, M.D.; data curation, A.F. and M.D.; writing—original draft preparation, G.M., M.D. and C.C.; writing—review and editing, A.F.; visualization, C.C. and G.M.; supervision, G.M.; project administration, C.C. All authors have read and agreed to the published version of the manuscript.

Funding: This research received no external funding.

Institutional Review Board Statement: The study was carried out in accordance with the ethical standards laid down in the 1975 Declaration of Helsinki. The Institutional Review Board of the Department of Human Sciences, Society, and Health of the University of Cassino and Lazio Meridionale approved this study (approval No.: 9947; date: 14 April 2021).

Informed Consent Statement: Informed consent was obtained from all subjects involved in the study.

Data Availability Statement: The data of this study are available upon request to the corresponding author at the address gabriele.mascherini@unifi.it.

Conflicts of Interest: The authors declare no conflict of interest.

References

1. Italian Ministry of Health. COVID-19—Situazione in Italia. Available online: http://www.salute.gov.it/portale/nuovocoronavirus/dettaglioContenutiNuovoCoronavirus.jsp?lingua=italiano&id=5351&area=nuovoCoronavirus&menu=vuoto (accessed on 30 July 2021).
2. Tomovic, M.; Krzman, L. Sport and exercise participation in time of COVID-19—A narrative review of medical and health perspective. *Transl. Sport. Med.* **2021**, *4*, 159–162. [CrossRef] [PubMed]
3. Du, Y.; Liu, B.; Sun, Y.; Snetselaar, L.G.; Wallace, R.B.; Bao, W. Trends in Adherence to the Physical Activity Guidelines for Americans for Aerobic Activity and Time Spent on Sedentary Behavior Among US Adults, 2007 to 2016. *JAMA Netw. Open* **2019**, *2*, e197597. [CrossRef] [PubMed]
4. Bull, F.C.; Al-Ansari, S.S.; Biddle, S.; Borodulin, K.; Buman, M.P.; Cardon, G.; Carty, C.; Chaput, J.P.; Chastin, S.; Chou, R.; et al. World Health Organization 2020 guidelines on physical activity and sedentary behaviour. *Br. J. Sport. Med.* **2020**, *54*, 1451–1462. [CrossRef] [PubMed]
5. Oliveira Neto, L.; Elsangedy, H.M.; Tavares, V.D.O.; Teixeira, C.V.L.S.; Behm, D.G.; Da Silva-Grigoletto, M.E. TreineEmCasa—Treinamento físico em casa durante a pandemia do COVID-19 (SARS-CoV-2): Abordagem fisiológica e comportamental. *Rev. Bras. Fisiol. Exerc.* **2020**, *19* (Suppl. 2), S9–S19. [CrossRef]
6. McNamara, J.M.; Swalm, R.L.; Stearne, D.J.; Covassin, T.M. Online weight training. *J. Strength Cond. Res.* **2008**, *22*, 1164–1168. [CrossRef]
7. Winters-Stone, K.M.; Boisvert, C.; Li, F.; Lyons, K.S.; Beer, T.M.; Mitri, Z.; Meyers, G.; Eckstrom, E.; Campbell, K.L. Delivering exercise medicine to cancer survivors: Has COVID-19 shifted the landscape for how and who can be reached with supervised group exercise? *Support. Care Cancer* **2022**, *30*, 1903–1906. [CrossRef]
8. Matysiak, M.; Siger, M.; Walczak, A.; Ciach, A.; Jonakowski, M.; Stasiołek, M. The influence of COVID-19 pandemic lockdown on the physical activity of people with multiple sclerosis. The role of online training. *Mult. Scler. Relat. Disord.* **2022**, *63*, 103843. [CrossRef]
9. Thompson, W.R. Worldwide Survey of Fitness Trends for 2022. *ACSM's Health Fit. J.* **2022**, *26*, 11–20. [CrossRef]
10. da Cunha de Sá-Caputo, D.; Taiar, R.; Seixas, A.; Sanudo, B.; Sonza, A.; Bernardo-Filho, M. A Proposal of Physical Performance Tests Adapted as Home Workout Options during the COVID-19 Pandemic. *Appl. Sci.* **2020**, *10*, 4755. [CrossRef]
11. WHO. *Waist Circumference and Waist-Hip Ratio: Report of a WHO Expert Consultation*; World Health Organization: Geneva, Switzerland, 2011. Available online: https://www.who.int/publications/i/item/9789241501491 (accessed on 30 July 2021).
12. American College of Sports Medicine Position Stand. The recommended quantity and quality of exercise for developing and maintaining cardiorespiratory and muscular fitness, and flexibility in healthy adults. *Med. Sci. Sports Exerc.* **1998**, *30*, 975–991. [CrossRef]
13. Gibson, A.L.; Wagner, D.; Heyward, H.V. Assessing flexibility. In *Advanced Fitness Assessment and Exercise Prescription*; Human Kinetics: Champaign, IL, USA, 2010; pp. 274–275.
14. Jain, N.B.; Wilcox, R.B.; Katz, J.N.; Higgins, L.D. Clinical examination of the rotator cuff. *PM&R* **2013**, *5*, 45–56. [CrossRef]
15. Hildebrandt, C.; Müller, L.; Zisch, B.; Huber, R.; Fink, C.; Raschner, C. Functional assessments for decision-making regarding return to sports following ACL reconstruction. Part I: Development of a new test battery. *Knee Surg. Sport. Traumatol. Arthrosc.* **2015**, *23*, 1273–1281. [CrossRef] [PubMed]
16. Roberts, H.C.; Denison, H.J.; Martin, H.J.; Patel, H.P.; Syddall, H.; Cooper, C.; Sayer, A.A. A review of the measurement of grip strength in clinical and epidemiological studies: Towards a standardised approach. *Age Ageing* **2011**, *40*, 423–429. [CrossRef] [PubMed]
17. Canadian Society for Exercise Physiology. *The Canadian Physical Activity, Fitness and Lifestyle Approach: CSEP-Health and Fitness Program's Health-Related Appraisal and Counseling Strategy*, 3rd ed.; Canadian Society for Exercise Physiology: Ottawa, ON, Canada, 2003; pp. 7–40.
18. Byrne, J.M.; Bishop, N.S.; Caines, A.M.; Crane, K.A.; Feaver, A.M.; Pearcey, G.E. Effect of using a suspension training system on muscle activation during the performance of a front plank exercise. *J. Strength Cond. Res.* **2014**, *28*, 3049–3055. [CrossRef] [PubMed]
19. Reimers, A.K.; Knapp, G.; Reimers, C.D. Effects of Exercise on the Resting Heart Rate: A Systematic Review and Meta-Analysis of Interventional Studies. *J. Clin. Med.* **2018**, *7*, 503. [CrossRef]
20. Piquet, L.; Dalmay, F.; Ayoub, J.; Vandroux, J.C.; Menier, R.; Antonini, M.T.; Pourcelot, L. Study of blood flow parameters measured in femoral artery after exercise: Correlation with maximum oxygen uptake. *Ultrasound Med. Biol.* **2000**, *26*, 1001–1007. [CrossRef]
21. Mannocci, A.; Masala, D.; Mei, D.; Tribuzio, A.M.; Villari, P.; LA Torre, G. International Physical Activity Questionnaire for Adolescents (IPAQ A): Reliability of an Italian version. *Minerva Pediatr.* **2021**, *73*, 383–390. [CrossRef]
22. American College of Sports Medicine (Ed.) *ACSM's Guidelines for Exercise Testing and Prescription*, 10th ed.; Wolters Kluwer: Philadelphia, PA, USA, 2018.
23. Dijkstra, H.P.; Ergen, E.; Holtzhausen, L.; Beasley, I.; Alonso, J.M.; Geertsema, L.; Geertsema, C.; Nelis, S.; Ngai, A.; Stankovic, I.; et al. Remote assessment in sport and exercise medicine (SEM): A narrative review and teleSEM solutions for and beyond the COVID-19 pandemic. *Br. J. Sports Med.* **2020**, *54*, 1162–1167. [CrossRef]
24. Hwang, R.; Fan, T.; Bowe, R.; Louis, M.; Bertram, M.; Morris, N.R.; Adsett, J. Home-based and remote functional exercise testing in cardiac conditions, during the COVID-19 pandemic and beyond: A systematic review. *Physiotherapy* **2022**, *115*, 27–35. [CrossRef]

25. D'Oliveira, A.; De Souza, L.C.; Langiano, E.; Falese, L.; Diotaiuti, P.; Vilarino, G.T.; Andrade, A. Home Physical Exercise Protocol for Older Adults, Applied Remotely During the COVID-19 Pandemic: Protocol for Randomized and Controlled Trial. *Front. Psychol.* **2022**, *13*, 828495. [CrossRef]
26. Kercher, V.M.; Kercher, K.; Bennion, T.; Levy, P.; Alexander, C.; Amaral, P.C.; Li, Y.M.; Liu, Y.; Wang, R.; Huang, H.Y.; et al. 2022 Fitness Trends from Around the Globe. *ACSM's Health Fit. J.* **2022**, *26*, 21–37. [CrossRef]
27. Cronshaw, S. Web workouts and consumer well-being: The role of digital-physical activity during the UK COVID-19 lockdown. *J. Consum. Aff.* **2022**, *56*, 449–464. [CrossRef] [PubMed]

sports

MDPI

Article

Applicability of an Immersive Virtual Reality Exercise Training System for Office Workers during Working Hours

Evlalia Touloudi *, Mary Hassandra, Evangelos Galanis, Marios Goudas and Yannis Theodorakis

School of Physical Education, Sport Science and Dietetics, Department of Physical Education and Sport Science, University of Thessaly, 42100 Trikala, Greece; maryhassandra@gmail.com (M.H.); egalanis@uth.gr (E.G.); mgoudas@pe.uth.gr (M.G.); theodorakis@uth.gr (Y.T.)
* Correspondence: elina.touloudi1995@gmail.com

Abstract: Virtual reality is a computer-generated simulation of a real or imaginary three-dimensional environment that has entered our lives, particularly for gaming. Lately, it has been permeating into many aspects of our everyday life, such as exercise. It is important to ascertain whether exercise in an immersive virtual reality environment can be accepted from employees and lead to positive outcomes for them. The aim of this exploratory study was to examine the acceptance, future adoption, interest/enjoyment and usability of an immersive virtual reality system for exercise training by office workers during breaks within their working hours. A total of 40 female employees participated in the study with a mean age of 42.58 years (SD 10.77). Participants were requested to complete two sequential 15-min dual task cycling sessions corresponding to two experimental conditions. The first, condition A, involved cycling in a virtual environment, wearing a virtual reality head mounted display, and responding to cognitive tasks by answering multiple choice questions—on a screen, using a joystick. The second, condition B, involved cycling on a static bicycle and simultaneously responding to cognitive tasks by answering multiple choice questions in a real environment. After completion of the two conditions, participants responded to a series of scales regarding each of the experimental conditions and to a semi-structured interview. The results showed that participants noted a significant preference for the immersive virtual reality exercise, condition A, compared to condition B (bike only); and their acceptance, interest/enjoyment, usability and intention for future use were high. The qualitative data showed increased intention for future use, feelings of control and presence and most of the participants did not encounter any difficulties or require extra help to understand the immersive virtual reality system. Overall, exercising during working hours with an immersive virtual reality exercise system was well perceived by office workers and applicable. However, the effects of the immersive virtual reality training system on physical and mental health and the employees' adherence to the exercise program should be tested with a longer intervention program.

Keywords: VR exercise system; office workers; sedentary; acceptability; usability; exercise

check for updates

Citation: Touloudi, E.; Hassandra, M.; Galanis, E.; Goudas, M.; Theodorakis, Y. Applicability of an Immersive Virtual Reality Exercise Training System for Office Workers during Working Hours. *Sports* **2022**, *10*, 104. https://doi.org/10.3390/sports10070104

Academic Editors: Rodrigo Zacca, Flávio Antônio de Souza Castro and Rui Miguel Simões de Azevedo

Received: 11 May 2022
Accepted: 27 June 2022
Published: 29 June 2022

1. Introduction

According to the World Health Organization (WHO), a great percentage of adults do not regularly meet the recommendations of physical activity needed to maintain or improve their health. This has as a result of poor health outcomes such as all-cause mortality, cardiovascular disease and type-2 diabetes, but also poor mental health and low general wellbeing. A strong recommendation is to limit the amount of sedentary time by replacing it with any intensity of physical activity [1]. The American College of Sports Medicine (ACSM) recommendation about physical activity is the daily goal of 30 min or more of moderate-intensity exercise. To attain the daily goal, individuals can have multiple 10-min or more bouts of exercise [2]. Many studies have examined the effects of short bouts of exercise versus one continuous bout per day. According to the results, multiple short sessions may enhance adherence to the exercise program, help weight loss and show similar improvements in cardiorespiratory function compared with long sessions [3]. In sedentary adults, multiple 10-min sessions of exercise may have positive effects in fitness and health benefits [4].

Specifically, according to a WHO recommendation, it is crucial for the health of sedentary employees to take a break from their job and exercise for 10–15 min. Regular exercise can lead to positive effects not only for the overall health of employees, but also to help them be more productive and decrease absenteeism. Employees can include multiple 10-min or more exercise bouts in their daily program to meet the WHO recommendations (>30 min per day) [1].

Currently, increased computer use has replaced manual work to a large extent [5]. The development of technology has offered new opportunities to all employees, including the disabled ones, for all stages of their lives and careers. Moreover, the advanced technology and remote work has reduced the risk of dangerous tasks, but led to difficulties in the determination of boundaries between private life and work, a lack of communication and a general psychosocial risk [6]. Additionally, employees face increased demands for performing faster and more complex tasks. A great percentage of employees report musculoskeletal disorders as the most serious work-related problem [5]. The costs of work-related illnesses are high, compared with the cost of wellbeing; therefore, mental and physical health-promoting programs are encouraged [6].

Many studies have been focusing on the efficacy of exercise in the workplace to improve employees' physical health, fitness, musculoskeletal problems, their productivity and mental health. Evidence shows that both endurance and muscle strengthening exercise can improve employees' health parameters such as body composition, motor, musculoskeletal and cardiorespiratory fitness and prevent health issues [7]. Undeniably, exercise in the workplace, especially for white-collar workers, can be beneficial for many parameters of their physical health, such as their fitness level, functional capacity, maximal strength, fat-free mass and musculoskeletal problems. This is very important because musculoskeletal pain has a negative effect on productivity and increases absenteeism [8]. Simultaneously, evidence shows that mental health can be improved because individuals have shown satisfaction towards their participation in training programs, improved their personal relationships [9] and increased perception about their health [10]. Additionally, implementing exercise programs in the workplace can lead to short-term financial savings by reducing healthcare costs and increasing the productivity of employees [11]. Work-related factors such as absenteeism, workability and work performance can also show positive effects from the employees' participation to exercise programs [12].

Immersive virtual reality (IVR) is a computer-generated simulation of a real or imaginary three-dimensional environment using a head mounted display and controllers connected with an application. According to researchers, individuals can exercise through the VR system and respond realistically to the virtual stimuli just like in real life [13]. The VR training is considered appropriate and can lead to positive effects in physiological, psychological and rehabilitation factors compared to conventional exercise [14]. Additionally, it is a feasible way of exercising and a powerful asset to increase physical and cognitive activity in both young and old individuals, since the elderly have reported similar experiences with younger adults [15]. So, a VR exercise program can be used for different ages, not only for aerobic, but also for muscle strengthening activities [16].

Recently, IVR has been widely used for many reasons, such as rehabilitation, training and prevention. For instance, exercise in a virtual environment has been used to help obese children to reach health recommendations [17], and healthy adults to improve balance and decrease the risk of falls [18]. Additionally, training in a virtual reality environment has been used by coaches for their athletes, especially in team ball sports, to fit them properly for any situation in the field [19].

A recent survey supports that since exercise is considered to have a therapeutic effect in chronic diseases, a combination of IVR and exercise can enhance the therapeutic results [20]. At the same time, IVR exercise programs have a positive effect on cognitive ability, attention and memory but also on non-cognitive factors such as depression and anxiety [21]. The combined exercise with cognitive training in an IVR environment has increasingly been studied, especially in patients with mild cognitive impairment symptoms. Hassandra et al. 2021 [22] employed an IVR system including physical and cognitive training for exercise patients with mild cognitive impairment. Their results showed that the system was acceptable, usable and tolerable for patients and that it could be used to promote their physical and cognitive health. Moreover, evidence shows that IVR exercise can lead to positive effects for mental health, particularly on depression and anxiety symptoms [23].

However, except for the positive effects of IVR exercise, it is important to take into consideration the adverse effects of IVR. Many people feel dizziness, discomfort or nausea while wearing the IVR headset and participating in IVR experiences. This situation is called cybersickness or simulator sickness [24].

Combined motor and cognitive dual-task is very common in our everyday life. According to MacPherson (2018) "dual-tasking is the ability to perform two tasks simultaneously and measures a component of executive function as participants are required to coordinate their attention to both tasks while they are being performed" [25]. According to studies, dual-task situations can improve diagnosis, prevention and management of cognitive impairment and falls [26]. Specifically, evidence shows that movement and posture include motor and cognitive components. According to Yogev-Seligmann et al. 2008 [27] during dual-tasking, one or both tasks are deteriorated, except for the situation when the cognitive demand is very low. However, if the motor skill is automated, the simultaneous execution of another task may not affect the task performance. In older adults, when simple tasks are less automatic, dual task costs may be increased. Hence, dual-task can become a useful clinical marker of cognitive impairment and fall risk, because it can worsen cognitive or motor disfunctions [26].

Virtual reality systems have recently started to be used in workplace environments for workers' stress reduction, relaxation, wellbeing improvement, work safety, work-related duties and thoughts distraction and have shown great expectation and support for VR interventions [28–30]. However, we have not detected a study that compared the perceptions of office workers regarding an IVR exercise system versus a static bicycle during working hours for training during working time. Therefore, the aim of this study was to examine the applicability of an IVR exercise system (compared to condition B: bike only) that allows office workers to exercise during their break within working hours. More specifically, we explored the employees' acceptance, future adoption, interest/enjoyment and usability of the IVR training system, because it can be used as a motivational tool for them to be more physically active. We expect that participants will prefer and enjoy the IVR training condition compared to the bike only condition. The novelties of this study are new data for the acceptance, usability, future adoption and interest/enjoyment of office workers towards the IVR training system during their working hours.

2. Materials and Methods

2.1. Participants and Setting

A total of 40 female office workers participated in this study. They ranged in age from 20 to 61 years old, with mean age 42.58 years (SD 10.77). Their educational level, habits of physical activity and technology use are presented in Table 1. The setting was administration offices in a public sector organization, in Volos, a city in central Greece. The administration offices employ 202 office workers, with 65 men and 137 women.

Table 1. Educational level, habits of physical activity and technology use.

Baseline Characteristic	*n*	%
Educational Level		
Secondary education	5	12.5%
Higher education	35	87.5%
Exercise program		
Yes	17	42.5%
No	23	57.5%
Physical Activity: Times Per Week		
One time per week	2	5%
Two times per week	4	10%
Three times per week	7	175%
Four times per week	2	5%
Five times per week	6	15%
Six Times Per Week	2	5%
Physical Activity: Hours Per Time		
45 min	5	12.5%
1 h	14	35%
1.5 h	4	10%
Phone Use		
Never	1	2.5%
1–2 h per day	17	42.5%
3–4 h per day	16	40%
5–6 h per day	4	10%
Up to 6 h per day	2	5%
Pc Use		
Never	2	5%
1–2 h per day	5	12.5%
3–4 h per day	4	10%
5–6 h per day	5	12.5%
Up to 6 h per day	24	60%
Video Game Use		
Never	35	87.5%
1–2 h per day	5	12.5%

2.2. Ethics Approval and Consent to Participate

Permission for the study was granted by the institution's ethics committee (approval number: 1829, 13 October 2021). The confidentiality of personal data will be assured with regulation (EU) 2016/679 (general data protection regulation). The participants were informed orally for the procedure, provided written information and signed the consent form. Additionally, they were encouraged to ask questions and received complete and detailed answers.

2.3. IVR System—Meta Quest (Devices and Application)

A cycle-ergometer (stationary seated bike type; Toorx, Chrono Line, BRX R 300) with Bluetooth capability has been used for the measurements, as an ideal solution for the participants to control their training conditions and minimize the risk of falls. The cycle-ergometer was connected with the IVR application, which includes motivational techniques to address the issue of low motivation to exercise and the IVR head mounted display and controllers. For the IVR projection, the Meta Quest 2 (Fcebook technologies, LCC, Hacker Way, Menlo Park, CA, USA) device was used (Figures 1 and 2). For example, participants could choose their exercise goal (exercise duration), have feedback for their training performance (cycling time, total distance, correct answers on cognitive exercises), task crafting (selection of music to enjoy while practicing) and self-monitoring (screen shows distance, time and speed data). The participants had the opportunity to choose the landscape in which they would like to practice (forest, beach, snowy landscape). In this study, all participants performed in the forest. The IVR system was developed by

ORAMA-VR and Biomechanical Solutions Engineering, on the basis of interviews with older people with mild cognitive impairment.

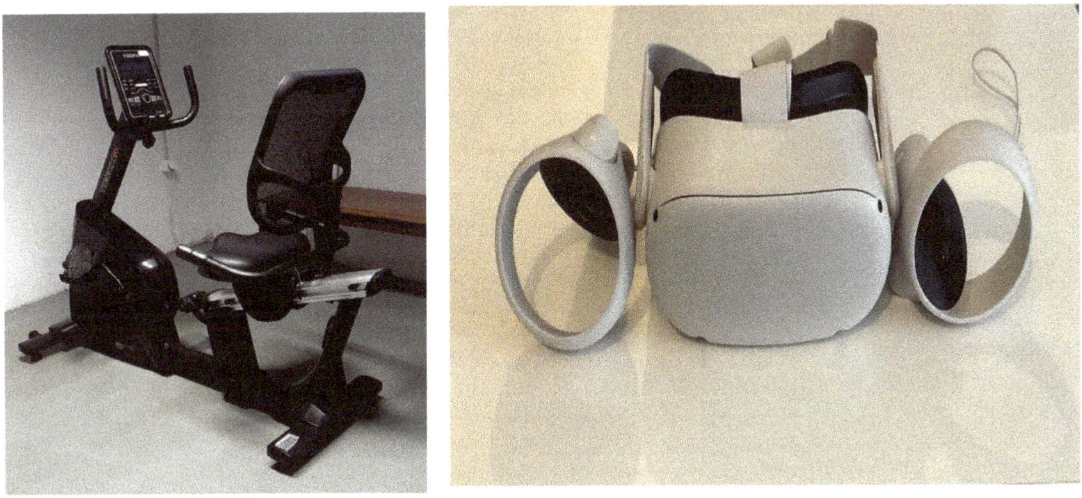

(a) (b)

Figure 1. (**a**) The cycle-ergometer; (**b**) the Meta Quest 2 head mounted display (HMD) and controllers.

Figure 2. Selection of exercise duration.

In the present study, the IVR training system was selected to be used by sedentary workers, as an exercise option during working hours. One of the reasons for the selection of the IVR bike is that it is a safe and attractive way to exercise in the workplace during the work break. Additionally, it is not time consuming, employees can choose their exercise duration by themselves and the cost is lower compared with a gym room in the workplace. Thus, in this way employees can take advantage of their work break to exercise.

The IVR training system used is called VRADA (VR exercise App for Dementia and Alzheimer's patients), Version 4.1. This IVR system (Version 3.7) was used with university students and patients with mild cognitive impairment as a dual task training aimed to promote their physical and cognitive health and assess its acceptability, tolerability and usability [22]. The results of this study were very encouraging, so it could be interesting to assess its applicability in different populations. The VRADA system has advantages that can meet the needs of employees, such as lack of time to exercise. Additionally, it is a very pleasant and attractive way to exercise and it might also be applicable to the employees, to have short breaks from work to exercise in their workplace, in order to maintain their physical health and avoid the negative effects from their sedentary work. (Supplementary Material S1—Video demo)

Firstly, the user must select the landscape for their training and then their cycling goal (training duration) within the IVR. As a selection mechanism, an IVR controller was used with a raycast allowing the user to choose an answer by pointing the ray at the button and pressing the trigger button on the controller.

Afterwards, they selected which word or phrase they would like to hear and repeat during their performance, such as "calmly", "I can", "I will do well", "Very nice" or no words. Then, the participants could begin their performance on the cycling-ergometer. They also had the opportunity to change the song they were listening to, from a list of preloaded tracks. At the end of the cycling program, they received feedback from the application about the correct and incorrect answers from the cognitive tasks and they could evaluate their experience through questions, e.g., "Are you tired today?", "Did you like the way you exercised today?", "How many animals did you see?", "How often did you repeat the word or phrase during the exercise?". In this study, cognitive tasks were used as a way of distraction to avoid possible boredom during cycling. There was not a suggestion to the participants to focus on the cognitive tasks.

2.4. Measures

2.4.1. Personal Innovativeness

A personal innovativeness questionnaire was used to find out users' beliefs about technology adoption decisions, with a Cronbach's α 0.77 [31]. The questionnaire included 4 items e.g., "I am the kind of person who looks forward to experimenting with new technologies" and participants were requested to answer through a 5-point Likert scale ranging from 1 (strongly disagree) to 5 (strongly agree).

2.4.2. Acceptance

Three factors were used to assess participants' acceptance for the VRADA app: 1. Perceived enjoyment; 2. Attitudes; 3. Intended future use.

1. *Perceived enjoyment,* with a Cronbach's α 0.94: 6 items were used to assess feelings of pleasure during exercise, e.g., "I really enjoyed exercising in the IVR environment" and participants were requested to answer through a 5-point Likert scale ranged from 1 (strongly disagree) to 5 (strongly agree) [32].
2. *Attitudes,* with a Cronbach's α 0.89: on the basis of guidelines from theory of planned behavior, 6 bipolar items were used to assess attitudes towards VRADA app e.g., "pleasant-unpleasant", "beautiful-ugly" and scored on a 7-point semantic differential scale [31].
3. *Intended future use,* with a Cronbach's α 0.94: on the basis of guidelines from the theory of planned behavior, 3 items were used to assess the extent to which an individual consciously wants to use the IVR system to exercise, e.g., "assuming I have access to the system, I intend to use it". Participants were requested to answer through a 5-point Likert scale ranged from 1 (strongly disagree) to 5 (strongly agree) [32,33].

2.4.3. Usability

Subjective usability was assessed with the system usability scale (SUS), a self-report questionnaire, with a Cronbach's α 0.71, which includes 10 items: 5 positive, e.g., "I think that I would like to use this system frequently" and 5 negative, e.g., "I found the system unnecessarily complex" [34]. Participants were requested to answer through a 5-point Likert scale ranging from 1 (strongly disagree) to 5 (strongly agree).

2.4.4. Preferences

To assess the preference for the exercise with or without the IVR system, 8 questions were used e.g., "Exercise was more pleasant ... ", "the numerical calculations were more fun ... "—"when exercising with or without the IVR system". Responses were dichotomous, scoring 0 (against IVR) or 1 (in favor of IVR). Respectively scoring was 0 (against the static bicycle) or 1 (in favor of the static bicycle). A composite score (total of the scores on the 8 questions) was computed for preference for the static bicycle and for preference for the IVR system. Thus, two preference scores, ranging from 0–8 were derived—One for the static bicycle and one for the IVR system.

2.4.5. Interest/Enjoyment

The intrinsic motivation inventory questionnaire, with a Cronbach's α 0.94, was used to assess the participants' interest/enjoyment for exercise with and without the IVR system with 6 items, e.g., "I enjoyed doing this activity very much", "This activity was fun to do". The questionnaire was used for both training conditions. Participants were requested to answer through a 5-point Likert scale ranged from 1 (strongly disagree) to 5 (strongly agree) [35,36].

2.4.6. IVR Equipment

Participants' perceptions of using the IVR system (headset and controller) were assessed by a questionnaire with 9 questions related to usability-pleasantness, with a Cronbach's α 0.64 (4 items e.g., "I felt comfortable using the head mounted display"), usability learning, with a Cronbach's α 0.58 (2 items, e.g., "It was easy to read the numerical questions") and tolerability, with a Cronbach's α 0.38 (3 items assessing dizziness, boredom and anxiety). Participants were requested to answer through a 5-point Likert scale ranged from 1 (strongly disagree) to 5 (strongly agree) [37].

2.4.7. Additional Assessments

Additional information was collected with a semi-structured interview. The semi-structured interview [38] was designed to collect qualitative data related to participants' subjective feelings and perceptions about the reasons for using VRADA, e.g., "Why would you use VRADA training system?", their expectations—2 items e.g., "Given the opportunity, would you be willing to use this training system regularly?", usability or utilization—5 items, e.g., "What difficulties did you encounter during the training session?", usability or learning, 2 items, e.g., "What exactly did you like most and least?", sense of presence or spatial presence, 2 items, e.g., "Did you feel you had control over the environment?", sense of presence or engagement, 2 items, e.g., "Did you get distracted during exercise? By what?", sense of presence or realism, e.g., "How did you find the environment, realistic or too artificial?" and tolerability, 2 items, e.g., "Did you feel bad during exercise? When and where exactly?". There were also 4 questions about the adoption of exercise with the IVR system to the participants' working environment, e.g., "How do you think using the IVR system to exercise can help you be more productive at work?", "If there were the VRADA system in your workplace, would you use it regularly? When? Why?". The interview guide can be found in Supplementary Material S2.

2.5. Procedure

Initially, participants were informed about the procedure, completed the consent form, adjusted their position on the cycle-ergometer and they were encouraged to ask questions about the procedure. The IVR training system was easy to use and instructions given were adopted from a previous study [22]. There were two different training conditions, condition A in a virtual environment and condition B in real environment, including a combination of cognitive and physical exercise. The use of IVR technology has lately been considered as a new approach to promoting physical activity and health behavior [39]. Early studies showed that VR bike exercise condition compared to the conditions without virtual reality significantly improved attendance and adherence (83.33%) over a 14-week intervention period [40]. A recent systematic review investigated the effectiveness of exercise-based VR training in improving physical activity and performance in a healthy population compared with traditional programs. Results showed a large effect on frequency of physical activity and a small to moderate effect on physical performance [41]. Additionally, higher enjoyment and self-efficacy and lower perceived exertion were reported after a VR exercise session compared to traditional exercise [42]. Wearing the VR headset while exercising can also make people remove themselves from their bodily sensations and allow them to exercise longer [43] and improve depression symptoms [44]. The VRADA exercise simulates exercise in a natural environment making it more attractive for short bouts of exercise during working time [45].

We compared the two conditions (A and B) to investigate which one is more preferable, enjoyable and interesting. Based on previous studies that compared similar exercise conditions, we expected that condition A would be preferable for office workers as well. Several previous studies compared exercise sessions with and without VR, for example, virtual reality rehabilitation versus conventional physical therapy in Parkinson's disease patients [46], VR cardiac rehabilitation program versus booklet or usual care [47], VR biking versus exergaming and traditional biking in college students [42] and VR biking versus traditional biking in college students [48]. Therefore, we considered it as important to follow a similar design for our comparator groups to provide richer content for our target group of office workers.

Participants completed each session in counterbalanced order. For the first two minutes, each individual had time to familiarize themselves with the cycle-ergometer, the headset and the controller and then they were requested to perform condition A: to cycle for 15 min in a virtual reality environment, using the VRADA system and answer 20 simple math calculations, displayed at the screen, using the controller. They could receive feedback about the cycling time, their speed, distance and correct answers on cognitive exercises on their screen. After completing this process, they had a 10-min break and then they started condition B: to cycle for 15 min in a real environment and answer 20 numerical questions orally. During the cycling process, each subject could receive feedback about the cycling time, their speed and distance on a connected computer and at the end they were informed about their score in cognitive questions. The cycling performance was at a constant speed ranged from 15 to 20 km/h. In every trial, the bike workload and speed were adjusted according to the exercise protocol, which was the same for both conditions. (Figure 3) Finally, after a 5-min rest, they completed a questionnaire and discussed the interview questions with the experimenter. The duration of the procedure was approximately 60 min.

2.6. Data Analysis

Quantitative data were analyzed using SPSS Statistics version 26 (IBM Corporation). Summary statistics were calculated for demographic characteristics and correlations among all examined variables were assessed using the Pearson coefficient. To examine the preference among exercise conditions, and identify statistical differences of interest/enjoyment between the two exercise conditions, paired-samples t-test was performed. Data of the semi-structured interview were analyzed using thematic analysis [49], which can offer rich insights into attitudes and beliefs by identifying patterns of ideas and responses. The topics of the discussion were based on previous studies, the main themes were predetermined (deductive approach). Second-order themes were analyzed using an inductive approach, allowing the data to determine subthemes.

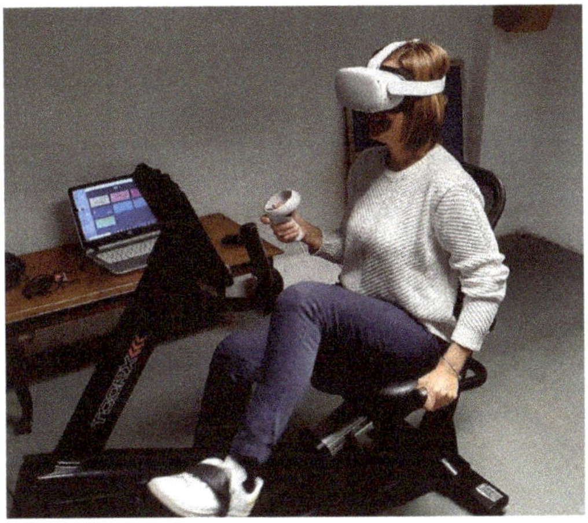

Figure 3. Participant performing an experiment.

3. Results

Table 2 presents descriptive statistics including the means, Cronbach's α and Pearson correlations for all the examined variables, whereas representation of data with boxplots are displayed in Supplementary Material S3.

Table 2. Main outcomes in the variables studied.

Variables	PI	PE	IFU	SUS	AT	IMI_A	IMI_B	PRE_A	PRE_B	UP	UL	MEAN	SD	Cronbach α
PI [a]												3.55/5	0.68	0.77
PE [b]	0.22											4.34/5	0.80	0.94
IFU [c]	0.25	0.80 **										4.02/5	0.97	0.94
SUS [d]	0.07	0.41 **	0.26									80.38/100	11.13	0.71
AT [e]	0.23	0.87 **	0.87 **	0.31 *								6.06/7	1.00	0.89
IMI_A [f]	0.27	0.83 **	0.68 **	0.42 **	0.82 **							4.21/5	0.78	0.94
IMI_B [g]	0.40 *	−0.00	−0.06	−0.01	−0.15	0.02						2.86/5	0.85	0.94
PRE_A [h]	0.07	0.68 **	0.72 **	0.18	0.73 **	0.64 **	−0.19					6.03/8	2.29	–
PRE_B [i]	−0.05	−0.69 **	−0.71 **	−0.20	−0.75 **	−0.66 **	0.21	−0.99 **				1.90/8	2.30	–
UP [j]	0.18	0.46 **	0.29	0.71 **	0.32 *	0.52 **	0.03	0.25	−0.25			4.18/5	0.56	0.64
UL [k]	0.17	0.09	0.06	0.52 **	0.00	0.15	0.20	0.08	−0.63	0.66 **		4.25/5	0.62	0.58
TOL [l]	−0.17	−0.21	−0.20	−0.17	−0.24	−0.36	−0.03	−0.32 *	−0.32 *	−0.31	−0.14	1.77/5	0.69	0.38

* Correlation is significant at the 0.05 level, ** Correlation is significant at the 0.01 level. [a] PI = personal innovativeness, [b] PE = perceived enjoyment, [c] IFU = intended future use, [d] SUS = system usability scale, [e] AT = attitudes, [f] IMI_A = interest/enjoyment for condition A (IVR) (intrinsic motivation inventory), [g] IMI_B = interest/enjoyment for condition B (intrinsic motivation inventory), [h] PRE_A = preference for condition A (IVR), [i] PRE_B = preference for condition B (bike only), [j] UP = usability—pleasantness, [k] UL = usability—learning, [l] TOL = tolerability.

3.1. Personal Innovativeness, SUS and Acceptance

The participants scored moderately to high on personal innovativeness and the SUS score was well above the acceptability threshold. Acceptance was assessed by three subfactors: perceived enjoyment, attitudes and intended future use. Participants scored highly on all three factors: perceived enjoyment, attitudes and intended future use.

3.2. Relationships between Variables

Personal innovativeness was low related to interest/enjoyment without IVR variable. Perceived enjoyment was moderately related to usability, preference for condition A and usability-pleasantness, strongly with intended future use, attitudes and interest/enjoyment for condition A and moderately negative to preference for condition B. Intended future use was moderately related to interest/enjoyment for condition A, strongly positive to attitudes and preference for condition A and strongly negative to preference for condition B. Usability had low relation with attitudes, moderate with interest/enjoyment for condition

A and usability-learning variable. Preference for condition A had a low negative relation with tolerability a moderate positive relation with interest/enjoyment for condition A and a strong positive relation with attitudes. Attitudes is strongly related with preference for the condition A and interest/enjoyment for the condition A, has low relation with usability-pleasantness and moderately negative related to preference for the condition B. The interest/enjoyment variable has moderate negative relation with preference for the condition B. Finally, the usability-pleasantness variable is moderately related to interest/enjoyment for condition A and usability-learning and strongly with usability.

3.3. Preferences and Interest/Enjoyment

A paired-samples t-test was conducted to compare the participants' preference between condition A and condition B. There was a significant difference in the scores favoring condition A (M = 6.03, SD = 2.29) in comparison to condition B (M = 1.90, SD = 2.30); t(39) = 5.81, $p < 0.001$.

A paired-samples t-test was performed to determine whether there was a statistically significant interest/enjoyment for either of the exercising conditions. There was a significant difference in the scores favoring condition A (M = 4.21, SD = 0.78) in comparison to condition B (M = 2.86, SD = 0.86); t(39) = 7.43, $p < 0.001$.

3.4. Semi-Structured Interview

The collected data from the semi-structured interview related to the participants' subjective feelings, expectations and perceptions about the VRADA's system usability and tolerability are presented in Table 3. A great percentage of participants reported that they would use this IVR system to exercise and that they did not encounter any difficulties. Most of them did not have any problems with the equipment (headset and joystick) but more than half of the participants noted a difficulty to aim the correct answers with the joystick. None of them needed extra help or time to understand the IVR system and most of them had feelings of control and presence in the IVR environment. Concerning the use of the IVR system in the workplace many participants reported that the system may have advantages for the employees like that it can help them to relax, to be more physically active and improve their mood and wellness. Finally, many of the participants reported that they find it realistic to exercise in their workplace with the VRADA system and the majority reported that they could use it systematically. However, some of the participants noted that they do not find it realistic to exercise with the IVR system during their working hours in the workplace. Most of them supported their answer by saying that they are working in high pressure and they do not have enough time for an active break.

Table 3. Subjective feelings, expectations and perceptions about the VRADA system, its usability and tolerability.

Main Themes	Subthemes	*n*	(%)
	Why would you use VRADA? Because:		
Reasons to use VRADA	• It is pleasant	25	62.5%
	• Time passed fast and delightfully	12	30%
	• It is interesting	7	17.5%
	• It is relaxing	6	15%
	Future personal use of the system		
	• Yes	24	60%
	• So-So	5	12.5%
	• No	11	27.5%
	Useful for other populations		
Expectations	• People who cannot exercise outdoors	12	30%
	• People who do not like to exercise, motivation	9	22.5%
	• Children and adolescents	9	22.5%
	• Disabilities	7	17.5%
	• Everybody	6	15%
	• People with limited free time	5	12.5%
	• People with mental disorders	5	12.5%

Table 3. *Cont.*

Main Themes	Subthemes	n	(%)
Usability or utilization	General difficulties		
	• None	27	67.5%
	• Combination of questions and searching for animals	4	10%
	• Questions placed very low	4	10%
	• To find animals	4	10%
	Technical issues		
	• To aim the correct answers with the joystick	22	55%
	• None	16	40%
	• The system was disconnected	2	5%
	Joystick use		
	• Ok	36	90%
	• Ineffective	2	5%
	• The button	1	2.5%
	IVR HMD use		
	• Ok	27	67.5%
	• Did not fit properly	9	22.5%
	• It was heavy	6	15%
Usability or learning	Need for extra help		
	• No	40	100%
	Need more time to understand the system		
	• No	40	100%
Usability or pleasantness	Most enjoyable parts		
	• The environment	32	80%
	• Music	17	42.5%
	Least enjoyable parts		
	• Repeated virtual parts	15	37.5%
	• Graphics	9	22.5%
	• Questions	8	20%
	Feel uncomfortable		
	• No	38	95%
	• I felt alone	1	2.5%
	• When I was searching for the animals	1	2.5%
Sense of presence or spatial presence	Control of the system		
	• Yes	33	82.5%
	• Not really	3	7.5%
	• Almost yes	2	5%
	• No	2	5%
	Feel part of the VR environment		
	• Yes	26	65%
	• Almost yes	9	22.5%
	• No	3	7.5%
	• Not really	2	5%
Sense of presence or engagement	Distraction of attention		
	• No	34	85%
	• Yes (e.g., when animals appeared)	6	15%
	Duration of the experience		
	• Enough	20	50%
	• It could be more	19	47.5%
	• It could be less	1	2.5%
Sense of presence or realism	VR environment was realistic or artificial		
	• Artificial	33	82.5%
	• Realistic	7	17.5%
Tolerability	Feel bad during training		
	• No	36	90%
	• Yes (anxiety, fear)	4	10%
Use in workplace	Advantages of the system for employees		
	• Help them to relax	17	42.5%
	• Help them to be physically active	10	25%
	• Improve mood and wellness	9	22.5%
	• Have a delightful break from work	8	20%
	• Strengthen physical and mental health	4	10%
	Help to be more productive		
	• Allow my mind to rest	13	32.5%
	• Calm down from stress and tension	12	30%
	• Have a delightful break from work	11	27.5%
	• It cannot help	3	7.5%
	How realistic is systematically exercise in workplace with the VR system		
	• Very realistic	16	40%
	• Not at all	10	25%
	• Not very realistic	9	22.5%
	• Quite realistic	5	12.5%
	Systematically use		
	• Yes	29	72.5%
	• No	11	27.5%

4. Discussion

The aim of this study was to examine the applicability of an immersive virtual reality system for exercise training by office workers during breaks within their working hours. We expected that participants will prefer and enjoy the IVR training condition compared to the bike only condition. The novelties of this study are new data for the acceptance, usability, future adoption and interest/enjoyment of office workers towards the IVR training system during their working hours. The acceptance of a new technology is mediated by various factors related to the user's psychology, the design of the system and the quality of the technology [50]. In addition, the effects of exercise in a virtual reality environment depends on the feeling of presence, the task and the IVR system [51].

The constructs used for this study are based on the concept of applicability. Preference is linked to an individual's affective response to exercise and is positively associated with exercise frequency, habit, vitality, enjoyment and wellbeing variables [52,53]. Therefore, exercise programs should be adjusted to the individual's preference in order to lead to increased levels of adherence. Enjoyment is considered an important physical activity determinant, since it reflects feelings of pleasure, liking and satisfaction and has shown positive associations with intention, increased attendance and adherence to the exercise program [52,54]. According to our results, participants reported higher enjoyment during IVR exercise compared to regular biking. Furthermore, physical activity is associated with age, sex, self-efficacy, health status and motivation [55]. Intrinsic motivation is defined as a form of motivation deriving from the innate needs for competence and self-determination, which, when satisfied, typically result in positive feelings of control [56]. IVR cycling can promote greater intrinsic motivation compared with traditional cycling, as our results suggest and this may motivate future physical activity participation [57]. Additionally, the link between task variety and enjoyment can lead to increased intrinsic motivation [58] and VR provides greater task variety than regular stationary biking. Interest is a form of motivation characterized by a focus on a certain object [56] and it is based on the appraisals that something is new, complex, uncertain and unexpected [58], such as while exercising with IVR technology. Attitude is a strong predictor of intention to participate and adhere to physical activity [59]; therefore, attitudes towards IVR exercise and personal innovativeness have a significant effect on the adoption of new technology [60]. Nevertheless, usability is a crucial factor that must be measured and conditions that hinder someone's interaction with the device must be taken into consideration [61]. Moreover, the greater the feeling of presence in the virtual reality that participants experience, the greater level of motivation they have [57]. All the above constructs are considered important contributing variables to assessing the applicability of the IVR exercise system that may facilitate office workers to exercise during their break of working time.

In this study, participants provided moderate to high scores on personal innovativeness, implying that they were familiarized with new technologies. They also indicated good acceptance of the IVR system as denoted by the relatively high scores in the three acceptance factors and in two usability facets. The high levels of usability indicated great acceptance and ease of use. As we hypothesized, a greater preference to condition A and high scores on interest/enjoyment in favor of IVR cycling have also been found. According to current literature, few studies compared VR exercise sessions with traditional exercise sessions and assessed preference and interest/enjoyment in various populations. The results for older people, university students and older people with mild cognitive impairment were very encouraging in favor of IVR exercise compared with traditional exercise [22,62]. Correlation results showed a positive and strong relationship between perceived enjoyment, intention for future use, attitudes and interest/enjoyment for the IVR training system. According to motivation theorists, when individuals perceive exercise as a pleasant activity, there are also positive associations with enjoyment, increased motivation, intention to continue exercise and persistence [63].

According to the results of the semi-structured interview, the majority of the participants reported that they preferred the VRADA exercising system in comparison to condition B. There were not many difficulties reported, except for the sensitivity of the joystick. All participants affirmed that they did not need extra help or time to understand the system and many of them reported that they would use it for exercising systematically. Additionally, no adverse effects were reported through the relevant items and semi-structured interview questions, for example for dizziness, nausea or discomfort, so the cybersickness effect was not detected in this study. Concerning the use of the IVR system in the workplace, although most of the participants noted that they found it realistic to use the system frequently during their working hours, there was a notable percentage reported that they did not find it realistic for many reasons, such as occupational stress and lack of time. In Greece, occupational stress is a frequent phenomenon because of economic instability. Most companies or organizations in the private and public sector are trying to manage this changing situation resulting in stressful conditions for employees [64]. Stress and quality of life are factors affected by Greece's economic crisis and high stress levels are related to low job satisfaction and quality of life [65]. In turn, this limits the possibility for scientifically proven interventions to be implemented in the workplace [64]. However, in the European Union (EU), many organizations are preparing and trying worksite health and wellness strategies to promote employees' health and prevent diseases. The concept of occupational safety and health differs across the EU countries, but it consists of some common elements such as healthy eating, smoking abstinence, mental health promotion, increased exercise and physical activity and health monitoring. According to the occupational safety and health campaign, a workforce that is healthy, motivated and well-qualified can become a key pillar of a healthy society with economic wellbeing [66].

Participants also reported many advantages of the IVR training system to the employees' productivity. Employees' productivity depends, among others, on mental health, physical health, job characteristics and support from the organization and colleagues [67]. Except for job characteristics, all of them can have positive effects from exercise in the workplace. According to a study, possible benefits can be offered for employees' productivity and absenteeism, from a workplace exercise intervention, by improving their musculoskeletal disorders and physical health [68]. Additionally, mental health can be improved from an exercise program in the workplace [69] as well as the personal relationships between employees [6]. The VRADA system has also been tested with different groups of adults with reported high acceptability, usability and tolerability indexes by students and older people with mild cognitive impairment [22]. All things considered, we can safely claim that the IVR app had a great applicability in this population, in this Greek city, in their working environment.

Strengths and Limitations

To the best of our knowledge, no previous study has examined the perceptions of office workers for an IVR exercise system using a static bicycle during working hours. The VRADA system includes motivational techniques to address the issue of low motivation to exercise, for example, goal setting and feedback. Additionally, the users have the opportunity to choose their pace, velocity, landscape and music. The system can be adjusted to individual needs and installed in the workplace to be easily accessible for the employees.

Despite these encouraging results about the acceptance, usability, interest/enjoyment and intention for future use, there are some limitations. Since the VRADA system has been tested for one session, long-term interventions have to be tested to examine if this way of training will remain enjoyable and interesting, as well as the training effect [22,70]. Moreover, future research could include assessments of exercise intensity using this application. Moderate intensity exercise is commonly considered beneficial for health, but we suggest a "self-selected intensity" plan. According to current research, when people select the exercise intensity by themselves, the sense of autonomy is promoted and that is related with more adaptive motivational and behavioral outcomes in a physical activity setting [71]. Specially, the frequency and intensity of IVR exercise programs during working hours can be increased to be more challenging for employees, but duration must remain low in order to be realistic as an option for exercise during working hours. Nevertheless, there were no adverse effects in this single 15-min session, but in long-term implementation, the exercise duration and frequency must be taken into consideration. According to recent literature, low duration and frequency of IVR exercise minimize the cybersickness effect [72]. We assessed cybersickness effects with three items included in the questionnaire and semi-structured interview. We suggest future studies should use the simulator sickness questionnaire (SSQ) so the data could be more comparable with other studies. According to a systematic review and meta-analysis, which investigated SSQ scores related with VR exercise in multiple studies, pooled participants' scores were relatively high and content was considered as a major contributing factor [73]. Additionally, the sample size was small and participants were women only, because more women volunteered to participate in this study and the offices where we addressed were female-dominated. Therefore, the emerged results cannot be generalized to the general population because male participants were not included. There is a need for more studies to examine the IVR system training in the workplace during working hours with a larger sample and gender balance. The aim of this study was to examine the applicability of the VRADA system, and not the efficacy and users' adherence in the exercise program. Future studies could evaluate the effects of regular training with the VRADA system on physical and mental health and quality of life.

5. Conclusions

Overall, exercising during working hours with an IVR exercise system was well perceived by office workers and applicable. Nevertheless, long-term intervention studies are required to explore the effectiveness on long-term benefits for several health and fitness dimensions and cost effectiveness of IVR exercise in sedentary workplace settings.

Supplementary Materials: The following supporting information can be downloaded at: https://www.mdpi.com/article/10.3390/sports10070104/s1, Supplementary Material S1: Video demo; Supplementary Material S2: Interview Guide; Supplementary Material S3: Boxplots.

Author Contributions: This manuscript was drafted by all authors. Conceptualization, Y.T. and E.T.; methodology, Y.T. and M.H.; software, E.G.; validation, M.H., Y.T., M.G., E.T. and E.G.; formal analysis, E.T. and M.G.; investigation, E.T.; resources, Y.T.; data curation, E.T.; writing—original draft preparation, E.T.; writing—review and editing, E.T., M.H. and M.G.; supervision, Y.T., M.H. and M.G.; project administration, E.T.; funding acquisition, Y.T. and M.H. All authors have read and agreed to the published version of the manuscript.

Funding: This research is co-funded by the European Regional Development Fund of the EU and Greek national funds through the Operational Program Competitiveness, Entrepreneurship, and Innovation, under the call RESEARCH–CREATE–INNOVATE (project code: T1EDK-01448). The funding agency played no role in the design or execution of the study, analysis or interpretation of the data, or decision to submit the results. Data are available upon request from YT.

Institutional Review Board Statement: The study was conducted in accordance with the Declaration of Helsinki, and approved by the Ethics Committee of the University of Thessaly (protocol code 1829, 13 October 2021).

Informed Consent Statement: Informed consent was obtained from all subjects involved in the study.

Data Availability Statement: Not applicable.

Acknowledgments: We thank all office workers for participating in this study.

Conflicts of Interest: The authors declare no conflict of interest.

References

1. World Health Organization (WHO). Guidelines on Physical Activity and Sedentary Behaviour. 2020. Available online: https://www.who.int/publications/i/item/9789240015128 (accessed on 15 January 2022).
2. Garber, C.E.; Blissmer, B.; Deschenes, M.R.; Franklin, B.A.; Lamonte, M.J.; Lee, I.M.; Nieman, D.C.; Swain, D.P. Quantity and quality of exercise for developing and maintaining cardiorespiratory, musculoskeletal, and neuromotor fitness in apparently healthy adults: Guidance for prescribing exercise. *Med. Sci. Sports Exerc.* **2011**, *43*, 1334–1359. [CrossRef] [PubMed]

3. Jakicic, J.M.; Wing, R.R.; Butler, B.A.; Robertson, R.J. Prescribing exercise in multiple short bouts versus one continuous bout: Effects on adherence, cardiorespiratory fitness and weight loss in overweight women. *Int. J. Obes. Relat. Metab. Disord.* **1995**, *19*, 893–901. Available online: https://pubmed.ncbi.nlm.nih.gov/8963358/ (accessed on 10 March 2022). [PubMed]
4. Lee, I.M.; Sesso, H.D.; Paffenbarger, R.S. Physical activity and coronary heart disease risk in men: Does the duration of exercise episodes predict risk? *Circulation* **2000**, *102*, 981–986. [CrossRef] [PubMed]
5. European Agency for Safety and Health at Work. Musculoskeletal Disorders and Psychosocial Risk Factors in the Workplace—Statistical Analysis of EU-Wide Survey Data. 2021. Available online: https://osha.europa.eu/en/publications/musculoskeletal-disorders-and-psychosocial-risk-factors-workplace-statistical-analysis-eu-wide-survey-data (accessed on 10 March 2022).
6. Directorate-General for Employment, Social Affairs and Inclusion (European Commission), European Commission. Communication from the Commission to the European Parliament, the Council, the European Economic and Social Committee and the Committee of the Regions EU Strategic Framework on Health and Safety at Work 2021–2027 Occupational Safety and Health in a Changing World of Work. 2021. Available online: https://op.europa.eu/en/publication-detail/-/publication/3eda1904-d817-11eb-895a-01aa75ed71a1/language-en (accessed on 12 March 2022).
7. Barranco-Ruiz, Y.; Villa-Gonzalez, E. Health related physical fitness benefits in sedentary women employees after an exercise intervention with Zumba fitness. *Int. J. Environ. Res. Public Health* **2020**, *17*, 2632. [CrossRef] [PubMed]
8. Karatrantou, K.; Gerodimos, V.; Manouras, N.; Vasilopoulou, T.; Melissopoulou, A.; Mesiakaris, A.F.; Theodorakis, Y. Health—promoting effects of a concurrent workplace training program in inactive office workers (healpworkers): A randomized controlled study. *Am. J. Health Promot.* **2020**, *1*, 11. [CrossRef]
9. Michishita, R.; Jiang, Y.; Ariyoshi, D.; Yoshida, M.; Moriyama, H.; Yamato, H. The practice of active rest by workplace units improves personal relationships, mental health, and physical activity among workers. *J. Occup. Health* **2017**, *59*, 122–130. [CrossRef]
10. Genin, M.P.; Degoutte, F.; Finaud, F.; Pereira, B.; Thivel, D.; Duclos, M. Effect of a 5-month worksite physical activity program on tertiary employees' overall health and fitness. *JOEM* **2017**, *59*, 2. [CrossRef]
11. Baicker, K.; Cutler, D.; Song, Z. Workplace wellness programs can generate savings. *Health Aff.* **2010**, *29*, 2. [CrossRef]
12. Grimani, A.; Aboagye, E.; Kwak, L. The effectiveness of workplace nutrition and physical activity interventions in improving productivity, work performance and workability: A systematic review. *BMC Public Health* **2019**, *19*, 1676. [CrossRef]
13. Tieri, G.; Morone, G.; Paolucci, S.; Iosa, M. Virtual reality in cognitive and motor rehabilitation: Facts, fiction and fallacies. *Expert Rev. Med. Devices* **2018**, *15*, 107–117. [CrossRef]
14. Qian, J.; McDonough, D.J.; Gao, Z. The Effectiveness of Virtual Reality Exercise on Individual's Physiological, Psychological and Rehabilitative Outcomes: A Systematic Review. *Int. J. Environ. Res. Public Health* **2020**, *17*, 4133. [CrossRef] [PubMed]
15. Sakhare, A.R.; Yang, V.; Stradford, J.; Tsang, I.; Ravichandran, R.; Pa, J. Cycling and Spatial Navigation in an enriched, immersive 3D virtual park environment: A feasibility study in younger and older adults. *Front. Aging Neurosci.* **2019**, *11*, 218. [CrossRef] [PubMed]
16. Feodoroff, B.; Konstantinidis, I.; Froböse, I. Effects of full body exergaming in virtual reality cardiovascular and muscular parameters: Cross-Sectional Experiment. *JMIR Serious Games* **2019**, *7*, e12324. [CrossRef] [PubMed]
17. Polechonski, J.; Nierwinska, K.; Kalita, B.; Wodarski, P. Can physical activity in immersive virtual reality be attractive and have sufficient intensity to meet health recommendations for obese children? A pilot study. *Int. J. Environ. Res. Public Health* **2020**, *17*, 8051. [CrossRef]
18. Prasertsakul, T.; Kaimuk, P.; Chinjenpradit, W.; Limroongreungrat, W.; Charoensku, W. The effect of virtual reality-based balance training on motor learning and postural control in healthy adults: A randomized preliminary study. *Bio Med. Eng.* **2018**, *17*, 124. [CrossRef]
19. Faure, C.; Limballe, A.; Bideau, B.; Kulpa, R. Virtual reality to assess and train team ball sports performance: A scoping review. *J. Sports Sci.* **2020**, *38*, 192–205. [CrossRef]
20. Silva Costa, M.T.; Pinheiro Vieira, L.; Barbosa, E.O.; Mendes Oliveira, L.; Maillot, P.; Ottero Vaghetti, C.A.; Carta, M.G.; Machado, S.; Gatica-Rojas, V.; Monteiro-Junior, R.S. Virtual Reality-Based Exercise with Exergames as Medicine in Different Contexts: A Short Review. *Clin. Pract. Epidemiol. Ment. Health* **2018**, *15*, 15–20. [CrossRef]
21. Ge, S.; Zhu, Z.; Wu, B.; McConnell, E.S. Technology-based cognitive training and rehabilitation interventions for individuals with mild cognitive impairment: A systematic review. *BMC Geriatr.* **2018**, *18*, 113. [CrossRef]
22. Hassandra, M.; Galanis, E.; Hatzigeorgiadis, A.; Goudas, M.; Mouzakidis, C.; Karathanasi, E.M.; Petridou, N.; Tsolaki, M.; Zikas, P.; Evangelou, G.; et al. A Virtual Reality App for Physical and Cognitive Training of Older People with Mild Cognitive Impairment: Mixed Methods Feasibility Study. *JMIR Serious Games* **2021**, *9*, 1. Available online: https://games.jmir.org/2021/1/e24170 (accessed on 20 December 2021). [CrossRef]
23. Zeng, N.; Pope, Z.; Lee, J.E.; Gao, Z. Virtual reality exercise for anxiety and depression: A preliminary review of current research in an emerging field. *J. Clin. Med.* **2018**, *7*, 42. [CrossRef]
24. Campo-Prieto, P.; Cancela, J.M.; Rodrguez-Fuentes, G. Immersive virtual reality as physical therapy in older adults: Present or future (systematic review). *Virtual Real.* **2021**, *25*, 801–817. [CrossRef]
25. MacPherson, S.E. Definition: Dual-tasking and multitasking. *Cortex A J. Devoted Study Nerv. Syst. Behav.* **2018**, *106*, 313–314. [CrossRef] [PubMed]

26. Bayot, M.; Dujardin, K.; Tard, C.; Defebvre, L.; Bonnet, C.T.; Allart, E.; Delval, A. The interaction between cognition and motor control: A theoretical framework for dual-task interference effects on posture, gait initiation, gait and turning. *Clin. Neurophysiol.* **2018**, *48*, 361–375. [CrossRef] [PubMed]
27. Yogev-Seligmann, G.; Hausdoff, J.M.; Giladi, N. The role of executive function and attention in gait. *Mov. Disord.* **2008**, *23*, 329–342. [CrossRef] [PubMed]
28. Naylor, M.; Morrison, B.; Ridout, B.; Campbell, A. Augmented Experiences: Investigating the Feasibility of Virtual Reality as Part of a Workplace Wellbeing Intervention. *Interact. Comput.* **2019**, *13*, 507–523. [CrossRef]
29. Ahmaniemi, T.; Lindholm, H.; Muller, K.; Taipalus, T. Virtual Reality experience as a stress recovery solution in workplace. In Proceedings of the 2017 IEEE life sciences conference (LSC), Sydney, NSW, Australia, 13–15 December 2017; pp. 206–209. [CrossRef]
30. Grassini, S.; Laumann, K. Evaluating the use of virtual reality in work safety: A literature review. In Proceedings of the 30th European Safety and Reliability Conference and the 15th Probablistic Safety Assessment and Management Conference, Venice, Italy, 1–5 November 2020.
31. Agarwal, R.; Prasad, J. A Conceptual and Operational Definition of Personal Innovativeness in the Domain of Information Technology. *Inf. Syst. Res.* **1998**, *9*, 204–215. [CrossRef]
32. Rasimah, C.M.Y.; Ahmad, A.; Zaman, H.B. Evaluation of user acceptance of mixed reality technology. *AJET* **2011**, *27*, 1369–1387. [CrossRef]
33. Ajzen, I. Constructing A Theory of Planned Behavior Questionnaire. 2006. Available online: https://people.umass.edu/aizen/ (accessed on 5 February 2022).
34. Brooke, J. SUS—A quick and dirty usability scale. In *Usability Evaluation in Industry*; Jordan, P.W., Thomas, B., McClelland, I.L., Weerdmeester, B., Eds.; Taylor & Francis Ltd.: Bristol, UK, 1996; pp. 189–194.
35. Goudas, M.; Biddle, S.; Fox, K. Perceived locus of causality, goal orientations, and perceived competence in school physical education classes. *Br. J. Educ. Psychol.* **1994**, *64*, 453–463. [CrossRef]
36. Goudas, M.; Dermitzaki, I.; Bagiatis, K. Predictors of student's intrinsic motivation in school physical education. *Eur. J. Psychol. Educ.* **2000**, *15*, 271–280. [CrossRef]
37. Mrakic-Sposta, S.; Di Santo, S.G.; Franchini, F.; Arlati, S.; Zangiacomi, A.; Greci, L.; Moretti, S.; Jesuthasan, N.; Marzorati, M.; Rizzo, G.; et al. Effects of Combined Physical and Cognitive Virtual Reality-Based Training on Cognitive Impairment and Oxidative Stress in MCI Patients: A Pilot Study. *Front. Aging Neurosci.* **2018**, *10*, 282. [CrossRef]
38. Pedroli, E.; Greci, L.; Colombo, D.; Serino, S.; Cipresso, P.; Arlati, S.; Mondellini, M.; Boillini, L.; Giussani, V.; Goulene, K.; et al. Characteristics, usability, and users experience of a system combining cognitive and physical therapy in a virtual environment: Positive bike. *Sensors* **2018**, *18*, 2343. [CrossRef] [PubMed]
39. Grace, A.S.J.; Fox, J. Immersive virtual environments, avatars, and agents for health. In *Encyclopedia of Health and Risk Message Design and Processing*; Oxford University: Oxford, UK, 2017; Volume 1.
40. Annesi, J.J.; Mazas, J. Effects of virtual reality-enhanced exercise equipment on adherence and exercise-induced feeling states. *Percept. Mot. Skills* **1997**, *85*, 835–844. [CrossRef] [PubMed]
41. Ng, Y.L.; Ma, F.; Ho, F.K.; Ip, P.; Fu, K.W. Effectiveness of virtual and augmented reality-enhanced exercise on physical activity, psychological outcomes, and physical performance: A systematic review and meta-analysis of randomized controlled trials. *Comput. Hum. Behav.* **2019**, *99*, 278–291. [CrossRef]
42. McDonough, D.J.; Pope, Z.C.; Zeng, N.; Liu, W.; Gao, Z. Comparison of college students' blood pressure, perceived exertion, and psychosocial outcomes during virtual reality, exergaming and traditional exercise: An exploratory study. *Games Health J.* **2020**, *9*, 290–296. [CrossRef] [PubMed]
43. McClure, C.; Schofield, D. Running virtual: The effect of virtual reality on exercise. *J. Hum. Sport Exerc.* **2020**, *15*, 861–870. [CrossRef]
44. Xu, W.; Liang, H.; Baghaei, N.; Ma, X.; Yu, K.; Meng, X.; Wen, S. Effects of an Immersive Virtual Reality Exergame on University Students' Anxiety, Depression, and Perceived Stress: Pilot Feasibility and Usability Study. *JMIR Serious Games* **2021**, *9*, 29330. [CrossRef]
45. Calogiuri, G.; Litleskare, S.; Fagerheim, K.A.; Rydgren, T.L.; Brambilla, E.; Thurston, M. Experiencing Nature through Immersive Virtual Environments: Environmental Perceptions, Physical Engagement, and Affective Responses during a Simulated Nature Walk. *Front. Psychol.* **2018**, *8*, 2321. [CrossRef]
46. Feng, H.; Li, C.; Liu, J.; Wang, L.; Ma, J.; Li, G.; Gan, L.; Shang, X.; Wu, Z. Virtual reality rehabilitation versus conventional physical therapy for improving balance and gait in parkinson's disease patients: A randomized controlled trial. *Med. Sci. Monit.* **2019**, *25*, 4186–4192. [CrossRef]
47. Vieira, Á.; Melo, C.; Machado, J.; Gabriel, J. Virtual reality exercise on a home-based phase III cardiac rehabilitation program, effect on executive function, quality of life and depression, anxiety and stress: A randomized controlled trial. *Disabil. Rehabil. Assist. Technol.* **2017**, *13*, 112–123. [CrossRef]
48. Zeng, N.; Pope, Z.; Gao, Z. Acute effect of virtual reality exercise bike games on college students' physiological and psychological outcomes. *Cyberpsychology Behav. Soc. Netw.* **2017**, *20*, 453–457. [CrossRef]
49. Braun, V.; Clarke, V. Using thematic analysis in psychology. *Qual. Res. Psychol.* **2006**, *3*, 77–101. [CrossRef]

50. Dillon, A.; Morris, M. User acceptance of new information technology—Theories and models. *Annu. Rev. Inf. Sci. Technol.* **1996**, *31*, 3–32.
51. Neumann, D.L.; Moffitt, R.L. Affective and Attentional States When Running in a Virtual Reality Environment. *Sports* **2018**, *6*, 71. [CrossRef] [PubMed]
52. Teixeira, D.S.; Rodrigues, F.; Machado, S.; Cid, L.; Monteiro, D. Did You Enjoy It? The Role of Intensity-Trait Preference/Tolerance in Basic Psychological Needs and Exercise Enjoyment. *Front. Psychol.* **2021**, *12*, 682480. [CrossRef]
53. Miller, B.M.; Bartholomew, J.B.; Springer, B.A. Post-exercise affect: The effect of mode preference. *J. Appl. Sport Psychol.* **2005**, *17*, 263–272. [CrossRef]
54. Jekauc, D. Enjoyment during Exercise Mediates the Effects of an Intervention on Exercise Adherence. *Psychology* **2015**, *6*, 48. [CrossRef]
55. Bauman, A.E.; Reis, R.S.; Sallis, J.F.; Wells, J.C.; Loos, R.J.F.; Martin, B.W.; Alkandari, J.R.; Andersen, L.B.; Blair, S.N.; Brownson, R.C.; et al. Correlates of physical activity: Why are some people physically active and others not? *Lancet* **2012**, *380*, 258–271. [CrossRef]
56. Heckhausen, J.; Heckhausen, H. *Motivation and Action*, 3rd ed.; Cambridge University Press: Cambridge, UK, 2018; pp. 583–584.
57. Liu, W.; Zeng, N.; Pope, Z.C.; McDonough, D.J.; Gao, Z. Acute effects of immercive virtual reality exercise on young adults' situational motivation. *J. Clincal Med.* **2019**, *8*, 1947. [CrossRef]
58. Dimmock, J.; Jackson, B.; Podlog, L.; Magaraggia, C. The effect of variety expectations on interest, enjoyment, and locus of causality in exercise. *Motiv. Emot.* **2013**, *37*, 146–153. [CrossRef]
59. Smith, R.A.; Biddle, S.J.H. Attitudes and exercise adherence: Test of the theories of reasoned action and planned behavior. *J. Sports Sci.* **1999**, *17*, 269–281. [CrossRef]
60. Jackson, A.J.; Yi, M.Y.; Park, S.J. An empirical test of three mediation models for the relationship between personal innovativeness and user acceptance of technology. *Inf. Manag.* **2013**, *50*, 154–161. [CrossRef]
61. Tuena, C.; Pedroli, E.; Trimarchi, P.D.; Gallucci, A.; Chiappini, M.; Goulene, K.; Gaggioli, A.; Riva, G.; Lattanzio, F.; Giunco, F.; et al. Usability issues of clinical and research applications of virtual reality in older people: A systematic review. *Front. Hum. Neurosci.* **2020**, *14*, 93. [CrossRef]
62. Htut, T.Z.C.; Hiengkaew, V.; Jalayondeja, C.; Vongsirinavarat, M. Effects of physical, virtual reality-based, and brain exercise on physical, cognition and preference in older persons: A randomized controlled trial. *Eur. Rev. Aging Phys. Act.* **2018**, *15*, 1–12. [CrossRef] [PubMed]
63. Rodrigues, F.; Teixeira, D.S.; Neiva, H.P.; Cid, L.; Monteiro, D. The bright and dark sides of motivation as predictors of enjoyment, intention and exercise persistence. *Scand. J. Med. Sci. Sports* **2020**, *30*, 787–800. [CrossRef] [PubMed]
64. Fotinatos-Ventouratos, R. The causes and consequences of organizational stress: The case of Greece. In *Organizational Stress around the World*; Sharma, K.A., Cooper, C.L., Pestonjee, D.M., Eds.; Routledge: New York, NY, USA, 2021; Available online: https://www.routledge.com/Organizational-Stress-Around-the-World-Research-and-Practice/Sharma-Cooper-Pestonjee/p/book/9780367263157 (accessed on 14 April 2022).
65. Alexopoulos, E.C.; Palatsidi, V.; Tigani, X.; Darviri, C. Exploring stress levels, job satisfaction, and quality of life in a sample of police officers in Greece. *Saf. Health Work* **2014**, *5*, 210–215. [CrossRef]
66. Guazzi, M.; Faggiano, P.; Mureddu, G.F.; Faden, G.; Niebauer, J.; Temporelli, P.L. Worksite health and wellness in the European Union. *Prog. Cardiovasc. Dis.* **2014**, *56*, 508–514. [CrossRef]
67. Stepanek, M.; Jahanshahi, K.; Millard, F. Individual, workplace and combined effects modeling of employee productivity loss. *J. Occup. Environ. Med.* **2019**, *61*, 469–478. [CrossRef]
68. Pereira, M.; Comans, T.; Sjøgaard, G.; Straker, L.; Melloh, M.; O'Leary, S.; Chen, X.; Johnston, V. The impact of workplace ergonomics and neck-specific exercise versus ergonomics and health promotion interventions on office worker productivity: A cluster-randomized trial. *Scand. J. Work Environ. Health* **2019**, *45*, 42–52. [CrossRef]
69. Proper, K.I.; Oostrom, S.H. The effectiveness of workplace health promotion interventions on physical and mental health outcomes—A systematic review of reviews. *Scand. J. Work Environ. Health* **2019**, *45*, 546–559. [CrossRef]
70. Mouat, B.; Smith, A.E.; Mellow, M.L.; Parfitt, G.; Smith, R.T.; Stanton, T.R. The Use of Virtual Reality to Influence Motivation, Affect, Enjoyment, and Engagement During Exercise: A Scoping Review. *Front. Virtual Real* **2020**, *1*, 564664. [CrossRef]
71. Zourbanos, N.; Hatzigeorgiadis, A.; Tsiami, A.; Tzatzaki, T.; Georgakouli, K.; Manthou, E.; Goudas, M.; Jamurtas, A.; Hatzoglou, C.; Chatzisarantis, N.; et al. An initial investigation of smokers' urges to smoke and their exercise intensity preference: A mixed-methods approach. *Cogent Med.* **2016**, *3*, 1149043. [CrossRef]
72. Kennedy, R.S.; Stanney, K.M.; Dunlap, W.P. Duration and exposure to virtual environments: Sickness and Curves During and Across Sessions. *Presence* **2000**, *9*, 463–472. [CrossRef]
73. Saredakis, D.; Szpak, A.; Birckhead, B.; Keage, H.A.D.; Rizzo, A.; Loetscher, T. Factors associated with virtual reality sickness in head-Mounted Displays: A systematic Review and Meta-analysis. *Front. Hum. Neurosci.* **2020**, *14*, 96. [CrossRef] [PubMed]

MDPI AG
Grosspeteranlage 5
4052 Basel
Switzerland
Tel.: +41 61 683 77 34

Sports Editorial Office
E-mail: sports@mdpi.com
www.mdpi.com/journal/sports

www.ingramcontent.com/pod-product-compliance
Lightning Source LLC
Chambersburg PA
CBHW062058220926
51850CB00010B/2807
9783725819867